新工科·高质量教材建设计划

Fundamentals of Mechanical Design Exercises Book

机械设计基础习题集

ZHANG PEIPEI
张培培 主编

电子科技大学出版社
University of Electronic Science and Technology of China Press
·成都·

图书在版编目(CIP)数据

机械设计基础习题集 = Fundamentals of Mechanical Design—Exercises Book / 张培培主编. 成都：成都电子科大出版社，2024. 8. -- ISBN 978-7-5770-1084-7

Ⅰ. TH122-44

中国国家版本馆 CIP 数据核字第 2024CJ5532 号

机械设计基础习题集 Fundamentals of Mechanical Design—Exercises Book

JIXIE SHEJI JICHU XITIJI

张培培　主编

策划编辑　刘　凡
责任编辑　刘　凡
责任校对　魏　彬
责任印制　段晓静

出版发行　电子科技大学出版社
　　　　　成都市一环路东一段159号电子信息产业大厦九楼　邮编 610051
主　　页　www.uestcp.com.cn
服务电话　028-83203399
邮购电话　028-83201495

印　　刷　成都市火炬印务有限公司
成品尺寸　185mm×260mm
印　　张　9.75
字　　数　252千字
版　　次　2024年8月第1版
印　　次　2024年8月第1次印刷
书　　号　ISBN 978-7-5770-1084-7
定　　价　39.00元

PREFACE 前言

The English-language textbook *Fundamentals of Mechanical Design* was published by the University of Electronic Science and Technology of China Press in June 2022, with the publication number 978-7-5647-9354-8.

To make the *Fundamentals of Mechanical Design* textbook more comprehensive and to unify the learning materials for Chinese students and international students studying in China, a bilingual supplementary study resource has been developed: the *Fundamentals of Mechanical Design—Exercises Book*.

Unlike common exercise books, this exercise book is a result of educational reform and is oriented towards new engineering courses. It aims to cultivate students' engineering abilities and innovative thinking. In addition to reinforcing and consolidating knowledge points (multiple-choice questions, fill-in-the-blank questions, and analysis and calculation problems), it also provides discussion topics and open questions. Many of the questions are sourced from Chinese materials and related international textbooks, broadening the scope of content use. Additionally, it includes two sets of simulated exam papers for self-assessment.

Upon completion, this bilingual exercise book, together with the *Fundamentals of Mechanical Design* textbook and the online courseware, forms a complete and newly reformed teaching resource for the course.

Target audience: Undergraduate and graduate students at higher education institutions that take this course in bilingual or English, as well as technical personnel engaged in mechanical design-related work in foreign enterprises.

The first draft of this exercise book has been strongly supported by graduate student Yue Feng. And it was used once in the batch 2020 of Robotics Engineering major and Industrial Engineering major at the School of Mechanical and Electrical Engineering, University of Electronic Science and Technology of China. The feedback from students on the draft was very constructive. I would like to express my heartfelt thanks.

Some pictures and content sources used in this book are not marked with names and sources on the web page. I would like to express my heartfelt thanks to the relevant authors.

This exercises book comes with answers and detailed explanations. Readers can scan the QR code below to download.

Due to the limited level of knowledge, the editor sincerely hopes that the readers will criticize and correct the errors and deficiencies of this book in the process of use. For comments on this book, please feel free to contact editor at peipei.zhang@uestc.edu.cn.

Editor

Chengdu, August 2024

前言 PREFACE

《Fundamentals of Mechanical Design（机械设计基础）》全英语教材于2022年6月在电子科技大学出版社出版。为了使得《Fundamentals of Mechanical Design（机械设计基础）》教材更完整，使中国学生和来华留学生学习资料更统一，编者又采用双语编写了此教材的配套学习资料《Fundamentals of Mechanical Design—Exercises book（机械设计基础习题集）》。

本习题集不同于常见的机械设计基础习题集，为编者教学改革成果的体现，以新工科建设为导向，以培养学生工程能力和创新思维为目标，除了知识点的巩固和强化（选择题、填空题、分析计算题），还提供了讨论题目和开放题目。其中题目多来自"机械设计基础"课程的中文资料和国外相关教材的资料，这使得本习题集的内容使用范围更广。另外，本习题集还配备了2套考试模拟试题，供学生自测使用。

本习题集出版后，与《Fundamentals of Mechanical Design（机械设计基础）》教材和在线课件形成了"机械设计基础"课程完整的全新改革的教学资料。

本书的使用对象是：高等院校采用双语或英语授此课的本专科生和来华留学生，在外企从事机械设计相关工作的技术人员。

本书初稿的撰写得到了电子科技大学机械与电气工程学院封岳同学的大力支持，同时本书初稿在电子科技大学机械与电气工程学院2020级机器人工程专业和2020级工业工程专业授课中使用了一次，同学们对初稿中存在的问题提出了很好的建议，在此表示衷心感谢。

本书使用的一些图片和内容来源网页上没有标注姓名和出处的资料，在此向相关作者表示衷心感谢。

本习题集配有答案和详细解释，请读者扫描下方二维码下载获取。

由于水平有限，编者恳切希望广大读者在使用过程中对本书的错误和欠妥之处批评指正。对本书的意见敬请联系：peipei.zhang@uestc.edu.cn

编　者

2024年08月于成都

CONTENTS 目 录

Chapter 1 Introduction
第1章 绪 论

1.1 Choice Questions

1. What is the kinematic unit of the machine? (　　)

A. Mechanism　　B. Component

C. Part　　D. Machine

2. What is the independent manufacturing unit? (　　)

A. Mechanism　　B. Component

C. Part　　D. Machine

3. What is the essential differences between a machine and a mechanism? (　　)

A. Whether it can perform effective mechanical work or convert mechanical energy.

B. Whether it is composed of many components.

C. Whether the components can move relative to each other.

4. Which of the following is a correct statement of the concept of component? (　　)

A. A component is a combination of machine parts.

B. A component is the assembly unit of a machine.

C. A component is the manufacturing unit of a machine.

D. A member is the kinematic unit of a machine.

5. Design is a (　　) process of making decisions .

A. single and non-circular　　B. multi-cycle

C. single-circle　　D. random

6. Which of the following is not the machine? (　　)

A. Wood Plane　　B. Pulley

C. Gear　　D. Lever

Wood Plane

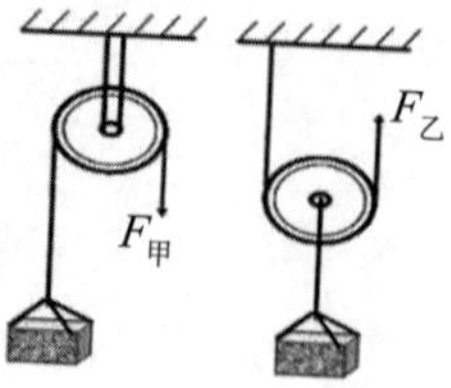

Pulley

Gear

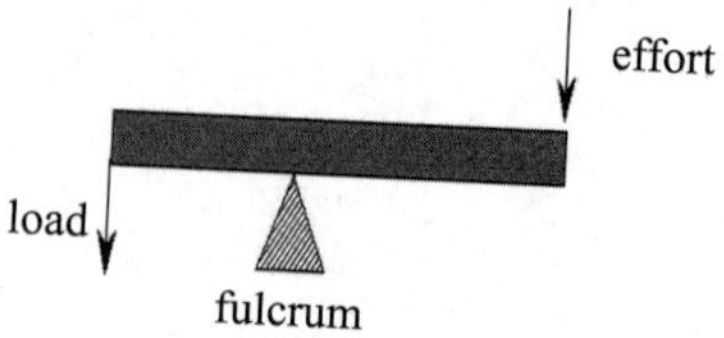

Lever

1.1 选择题

1. 组成机器的运动单元是什么？（　　）

A. 机构　　B. 构件

C. 零件　　D. 机器

2. 下列哪一项是独立的制造单元？（　　）

A. 机构　　B. 构件

C. 零件　　D. 机器

3. 机器与机构的本质区别是什么？（　　）

A. 是否能完成有用的机械功或转换机械能。

B. 是否由多个构件组合而成。

C. 各构件间能否产生相对运动。

4. 下列哪一项是构件概念的正确表述？（　　）

A. 构件是机器零件组合而成的。

B. 构件是机器的装配单元。

C. 构件是机器的制造单元 。

D. 构件是机器的运动单元。

5. 设计是一个（　　）决策的过程。

A. 单次不循环　　B. 多次循环

C. 单次循环　　D. 随机

6. 下列哪一项不是机器？（　　）

A. 木工刨　　B. 滑轮

C. 齿轮　　D. 杠杆

木工刨

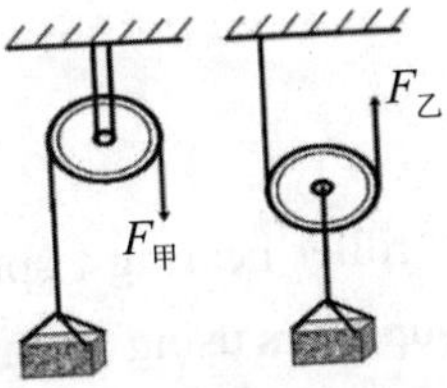

滑轮

齿轮

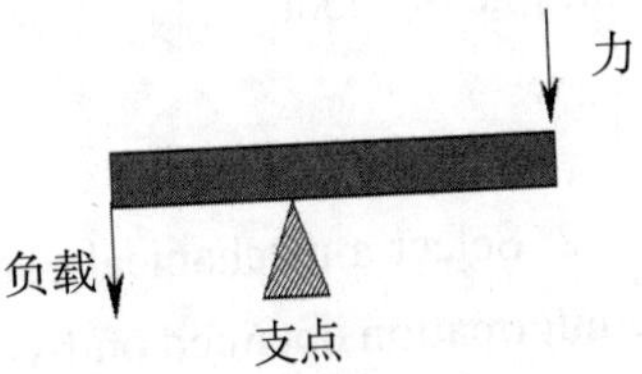

杠杆

1.2 Multiple Choice Questions

1. Which of the following are the machine? (　　)

A. Bicycle　　B. Machine tool

C. Watch　　D. Motor fan

2. Which are the design considerations? (　　)

A. Cost　　B. Reliability

C. Functionality　　D. Strength

1.3 Fill-in-the-Blanks Questions

1.The purpose of machine is to help people do things with some__________, so as to __________, __________ and make operation__________.

2. A component is the __________ unit of a machine. A part is the __________ unit of a machine.

3. The non-separating basic unit of the machine is the __________, which is the unit body of__________.

4. Design is an _______ and highly_______process. It is also the process of_______.

5.The full name of CAD is________________. The full name of CAM is ________________. The full name of CAE is________________.

1.4 Short-Answer Questions

1. Please explain design process and design considerations through an example such as bicycle or robot.

2. Select a mechanical component you know (roller bearings, springs, etc.), report on the information obtained on five manufacturers or suppliers using a search engine.

1.2 多选题

1. 下列哪些是机器？（　　）

A. 自行车　　B. 机床

C. 手表　　D. 电扇

2. 设计中的注意事项有哪些？（　　）

A. 成本　　B. 可靠性

C. 功能　　D. 强度

1.3 填空题

1. 机器的目的是帮助人们用一些__________做事情，从而节省__________和__________，使操作变得__________。

2. 构件是机器的__________单元；零件是机器的__________单元。

3. 机械中不可拆卸的基本单元为__________，它是__________的单元体。

4. 设计是一个______和______的______过程。

5. CAD是指__________，CAM是指__________，CAE是指__________。

1.4 简答题

1. 用自行车或机器人作为例子，分析设计过程和设计注意事项。

2. 选择一个机械零部件（如滚动轴承、弹承等），使用搜索引擎寻找全球或国内五个制造商或供应商的信息，给出查找详细信息的链接。

Chapter 2 Kinematics Fundamentals 1
第2章 运动学基本原理1

2.1 Choice Questions

1. How many degrees of freedom does a rigid body have in space? ( )
How many degrees of freedom does a rigid body have in plane? ( )

A. 3
B . 5
C. 7
D. 6

2. How many constraints are provided by the plane kinematic pairs? ( )

A. 1
B. 2
C. 1 or 2
D. 3

3. If the two components form a lower pairs, what is the contact form? ( )

A. Surface contact
B. Point or line contact
C. Point or surface contact
D. Line or surface contact

4. If the two components form a higher pairs, what is the contact form? ( )

A. Surface contact
B. Point or line contact
C. Point or surface contact
D. Line or surface contact

5. How many higher pairs exist in the mechanism shown below? ( )

A. 0
B. 1
C. 2
D. 3

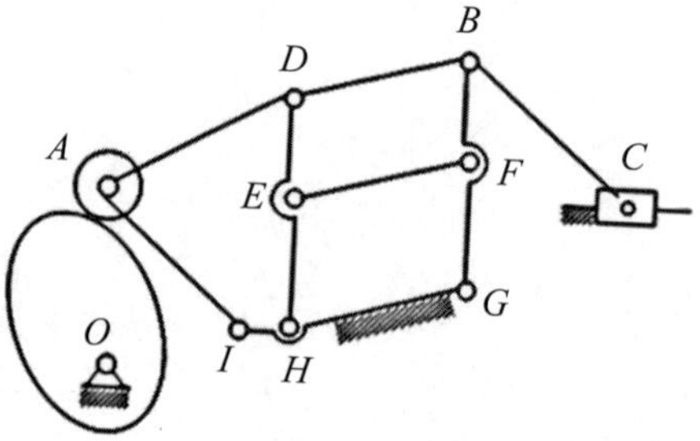

2.1 选择题

1. 一个构件在空间中的自由度为（　　），在平面上的自由度为（　　）。

A. 3　　B . 5

C. 7　　D. 6

2. 平面运动副所提供的约束为（　　）个。

A. 1　　B. 2

C. 1或2　　D. 3

3. 若两构件组成低副，则其接触形式为（　　）。

A. 面接触　　B. 点或线接触

C. 点或面接触　　D. 线或面接触

4. 若两构件组成高副，则其接触形式为（　　）。

A. 面接触　　B. 点或线接触

C. 点或面接触　　D. 线或面接触

5. 下图所示机构中存在几个高副？（　　）

A. 0　　B. 1

C. 2　　D. 3

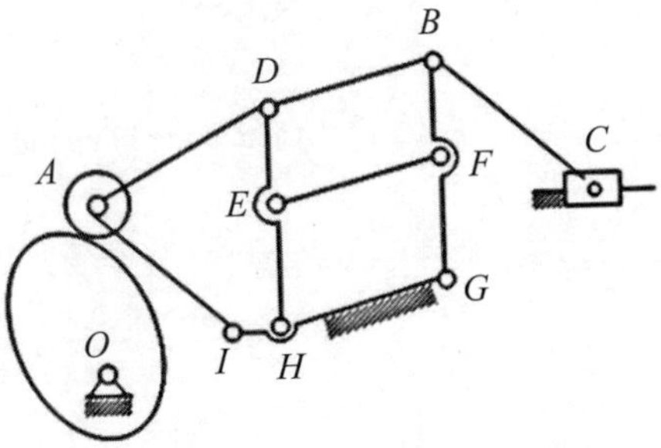

2.2 Fill-in-the-Blanks Questions

1.The system's DOF is equal to the________of________parameters which are needed to uniquely define its________in space at any instant of time.

2. The planar motion can be divided into________,________and________.

3. The kinematic pairs is a connection in which________components directly contact and can produce________.

4. Plane kinematic pairs can be divided into________and________, and lower pairs can be divided into________and________.

5. In the Crank Rocker mechanism shown below, ①, ②, ③ and ④ respectively are ________, ________, ________and________.

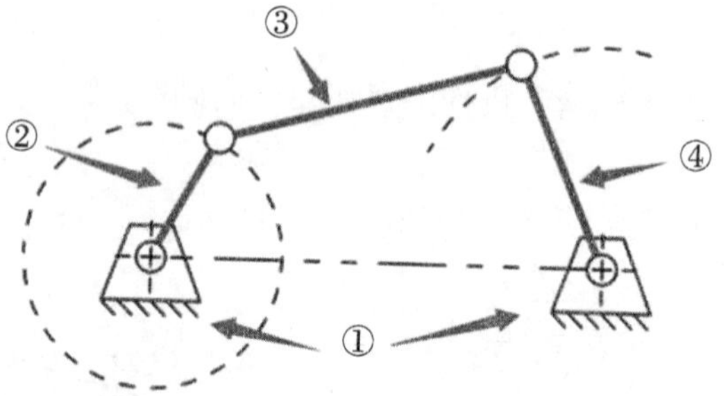

(The dotted line indicates the motion path)

2.2 填空题

1. 系统的自由度等于在任何时刻唯一定义其在空间中的__________所需的__________参数的__________。

2. 平面内运动的类型有：________、________和________。

3. 运动副是________（个）构件直接接触并能产生________的活动连接。

4. 平面运动副可分为________和________，低副又可分为________和________。

5. 下图所示曲柄摇杆机构中①、②、③、④分别是__________、__________、__________、__________。

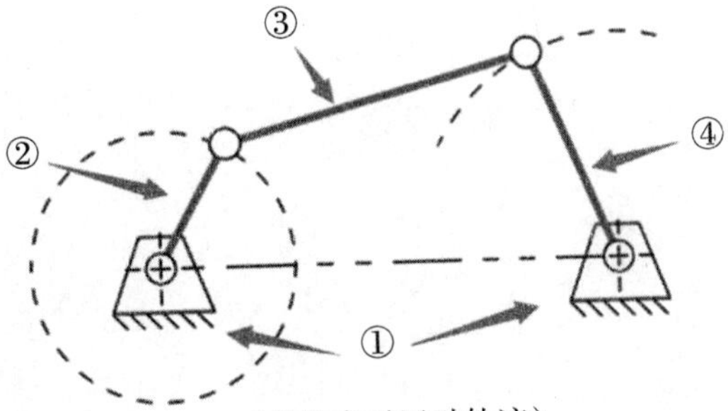

（虚线表示运动轨迹）

Chapter 3 Kinematics Fundamentals 2

第3章 运动学基本原理2

3.1 Choice Questions

1. The degrees of freedom of two planar mechanisms are equal to 1, and now a link with two hinges is used to connect them into a planar mechanism, the degree of freedom is equal to (　　).

A. 0

B. 1

C. 2

2. When calculating the degree of freedom of the mechanism, if the useless constraints are included, the degree of freedom of the mechanism will be (　　).

A. increased

B. decreased

C. unchanged

3. When calculating the degree of freedom of the mechanism, if the passive DOF is included, the degree of freedom of the mechanism will be (　　).

A. increased

B. decreased

C. unchanged

4. The condition for a mechanism to have a definite relative motion is that (　　).

A. the number of degrees of freedom of the mechanism is equal to the number of driving parts

B. the number of degrees of freedom of the mechanism is greater than the number of driving parts

C. the number of degrees of freedom of the mechanism is less than the number of driving parts

D. the number of degrees of freedom of the mechanism is greater than or equal to the number of driving parts

3.1 选择题

1. 有两个平面机构的自由度都等于1，现用一个带有两铰链的运动构件将它们连接成一个平面机构，则其自由度等于（　　）。

A. 0

B. 1

C. 2

2. 计算机构自由度时，若计入虚约束，则机构自由度会（　　）。

A. 增多

B. 减少

C. 不变

3. 计算机构自由度时，若计入局部自由度，则机构自由度会（　　）。

A. 增多

B. 减少

C. 不变

4. 机构具有确定相对运动的条件是（　　）。

A. 机构的自由度数目等于主动件数目

B. 机构的自由度数目大于主动件数目

C. 机构的自由度数目小于主动件数目

D. 机构的自由度数目大于等于主动件数目

3.2 Fill-in-the-Blanks Questions

1. The purpose of using useless constraints in mechanism is to improve the__________ condition and__________ condition of mechanism.

2. The condition that the relative motion of the mechanism is determined is that the number of driving parts__________the number of degrees of freedom of the mechanism.

3.The three considerations in determining degree of freedom:__________, __________, and__________.

4. If the number of drivers is__________ the degree of freedom for a mechanism, the mechanism will be broken. If the number of drivers is__________ the degree of freedom for a mechanism, the motion will be uncertain.

5. The formula for calculating the degree of freedom of planar mechanism is "__________". Where P_L is the number of__________ and P_H is the number of__________.

6. The number of useless constraints in the following mechanism is__________.

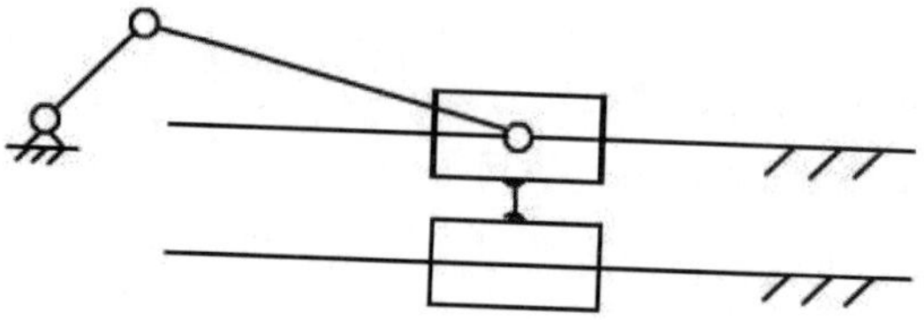

7. In order to have a definite motion of the following mechanism, the number of driver required is______.

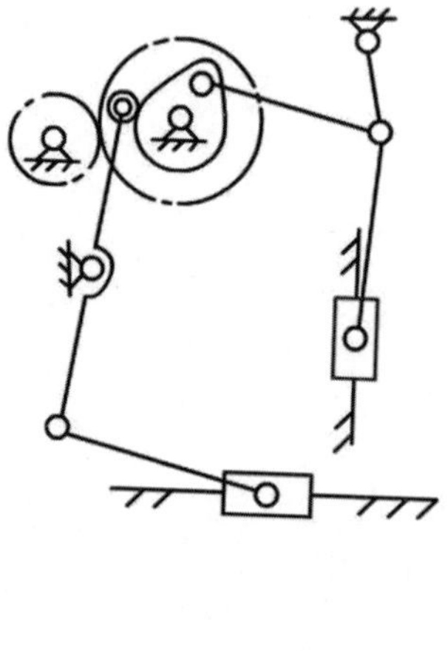

3.2 填空题

1. 在机构中采用虚约束的目的是改善机构的________和________。

2. 机构具有确定相对运动的条件是原动件的数目________机构的自由度数目。

3. 计算机构自由度时的三个注意事项是:________、________和________。

4. 如果原动件的数目________机构的自由度数目,则机构中最弱的构件将损坏。如果原动件数目________机构的自由度数目,机构运动不确定。

5. 计算平面机构自由度的公式为________。其中,P_L为________个数,P_H为________个数。

6. 图示机构中有________个虚约束。

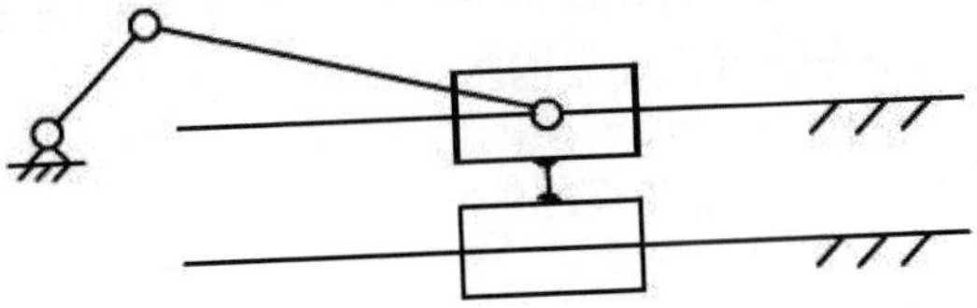

7. 图示机构要有确定运动,需要有________个原动件。

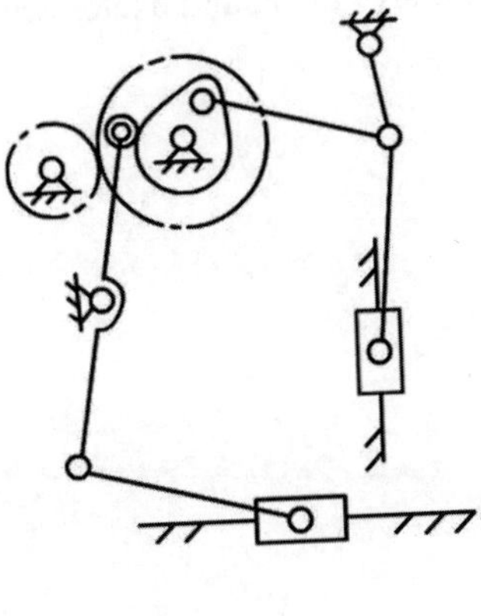

3.3 Analysis and Calculation Questions

1. Please state the conditions for determining the motion of a mechanism. And determine whether the mechanism motion in the figure below is certain.

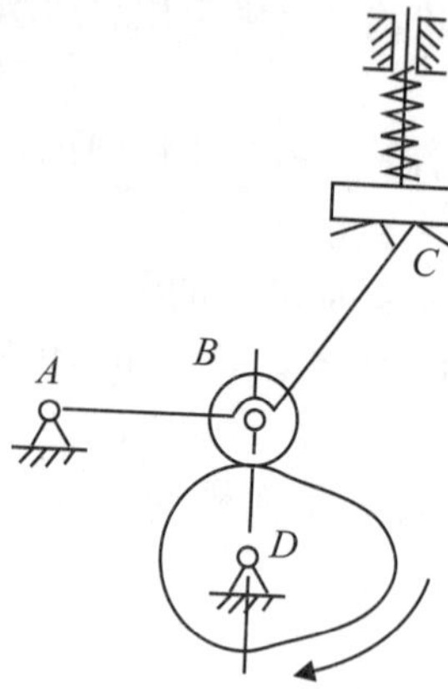

2. Find the degrees of freedom of the mechanism shown. Specify if there are multiple joints, passive DOF, and useless constraints.

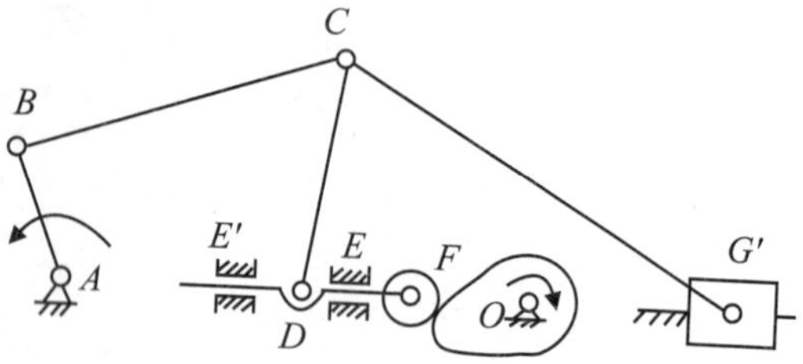

3. Find the degrees of freedom of the mechanism shown. Specify if there are multiple joints, passive DOF, and useless constraints.

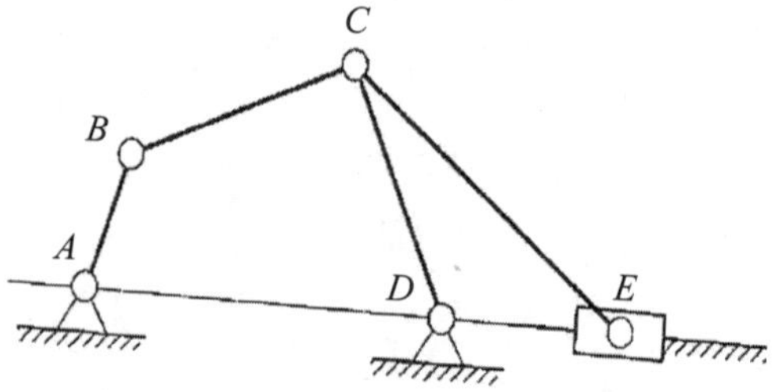

3.3 分析计算题

1. 说明图示机构具有确定相对运动的条件，并判断该机构运动是否确定。

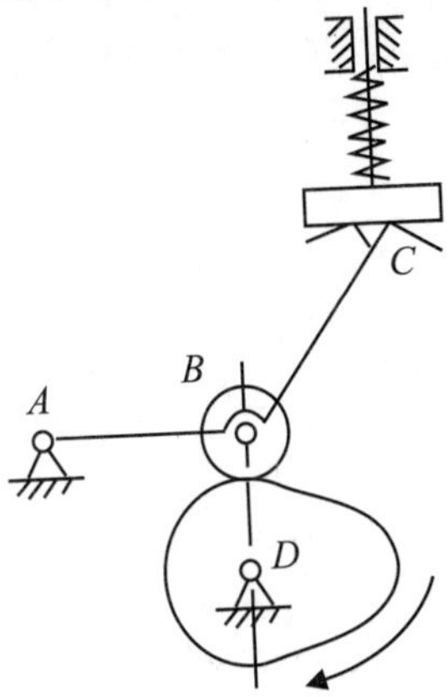

2. 计算图示机构的自由度，若含有复合铰链、局部自由度和虚约束请明确指出。

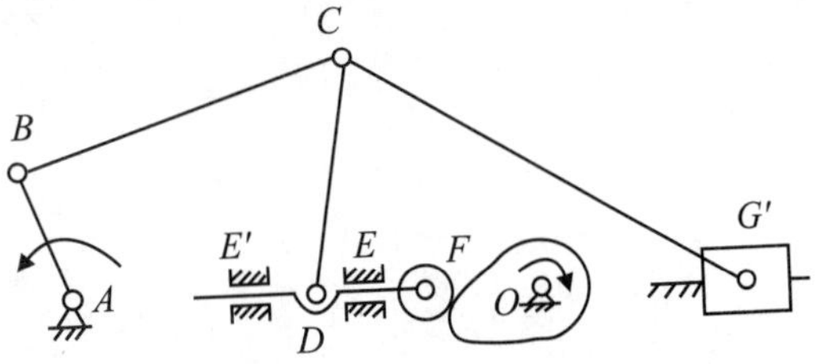

3. 计算图示机构的自由度，若含有复合铰链、局部自由度和虚约束请明确指出。

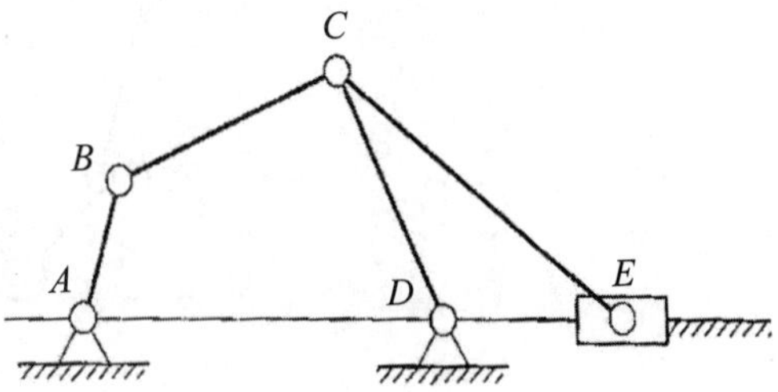

4. Find the degrees of freedom of the mechanism shown in the figure. Specify if there are multiple joints, passive DOF, and useless constraints.

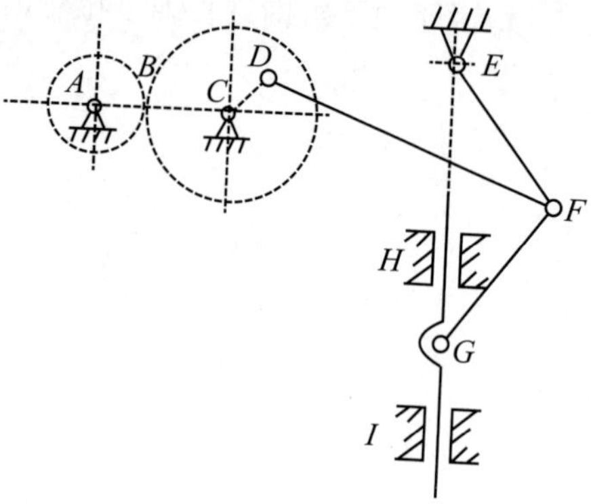

5. Find the degrees of freedom of the following mechanisms.

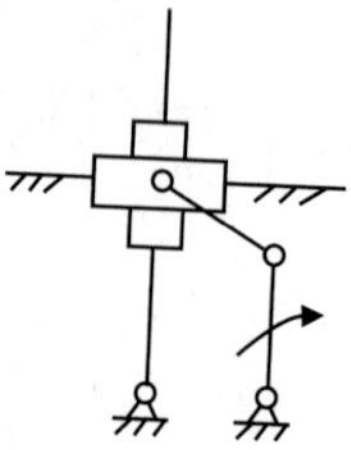

(a)

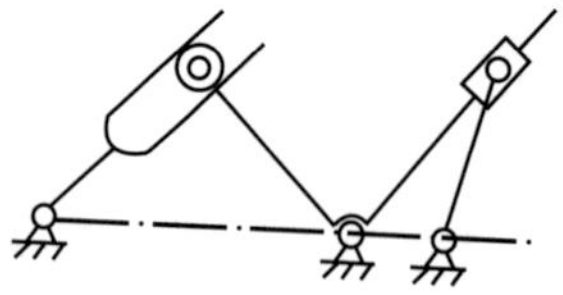

(b)

6. Find the degrees of freedom of the following mechanisms.

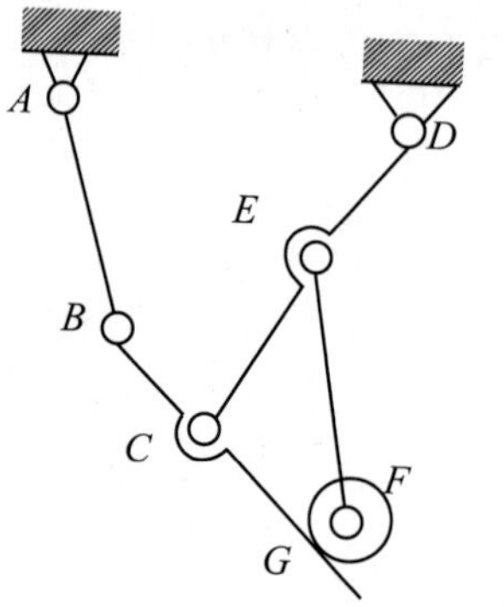

(a)

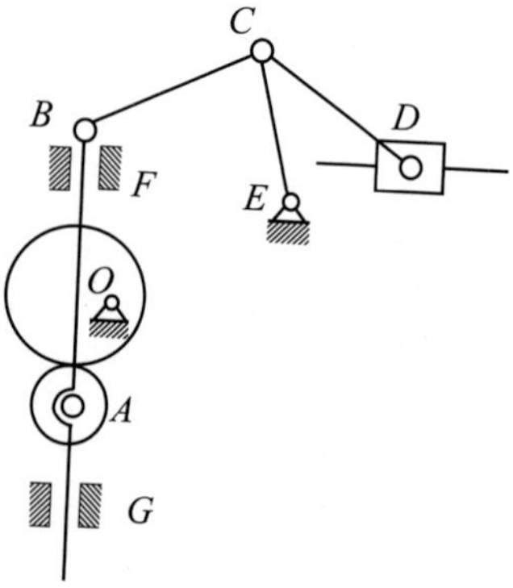

(b)

4. 计算图示机构的自由度，若含有复合铰链、局部自由度和虚约束请明确指出。

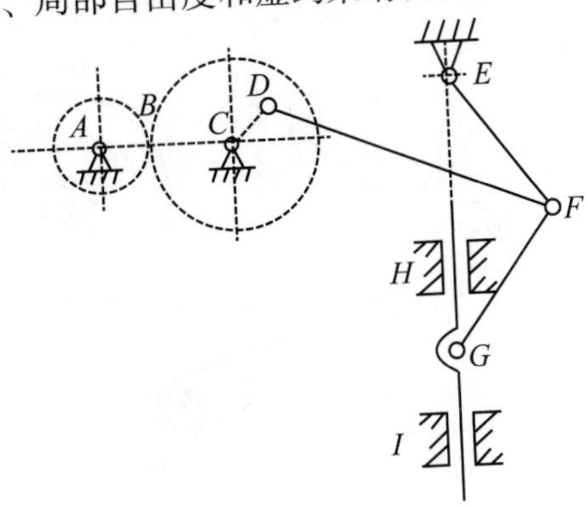

5. 计算下列机构的自由度。

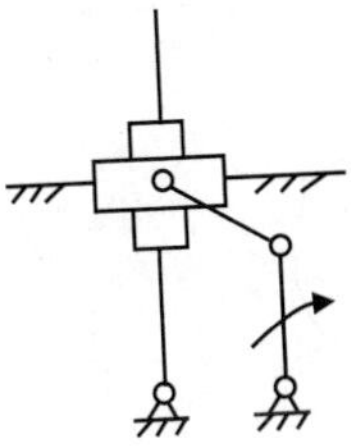

(a)

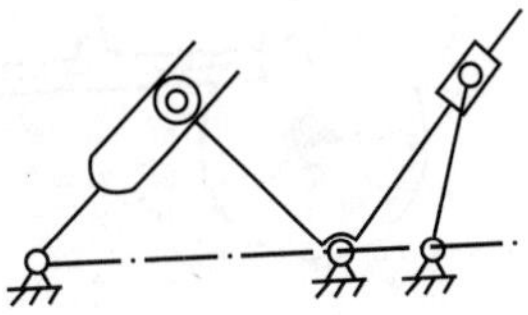

(b)

6. 计算下列机构的自由度。

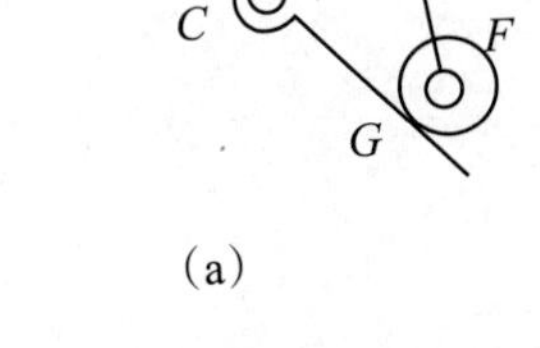

(a)

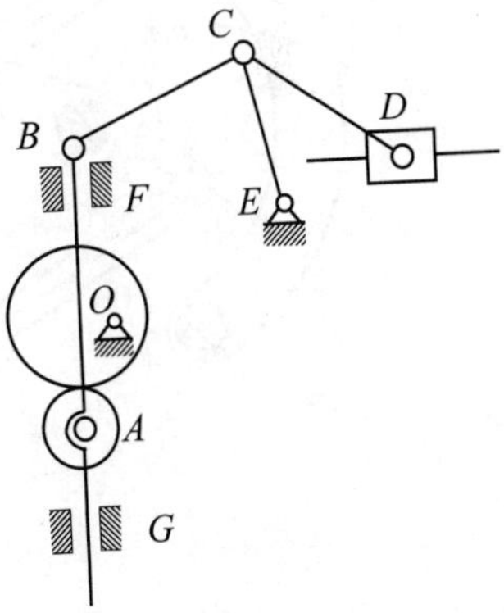

(b)

7. Find the degrees of freedom of the following mechanisms.

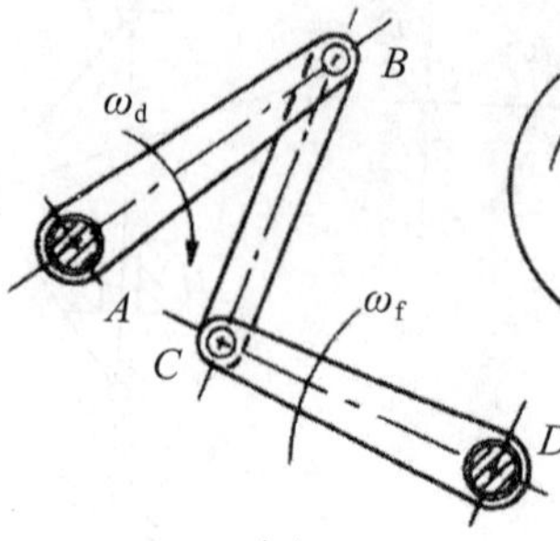

(a)

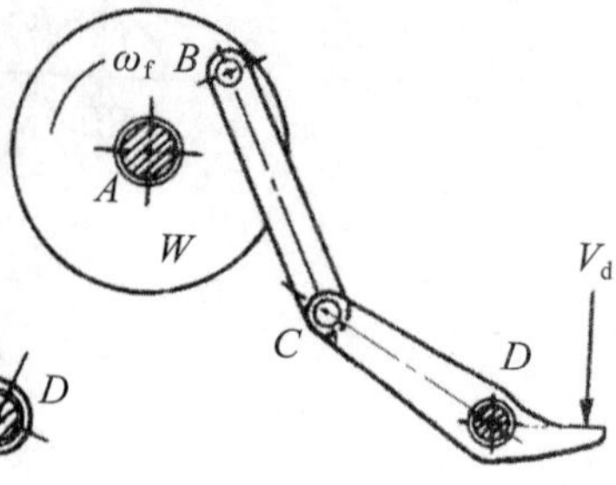

(b)

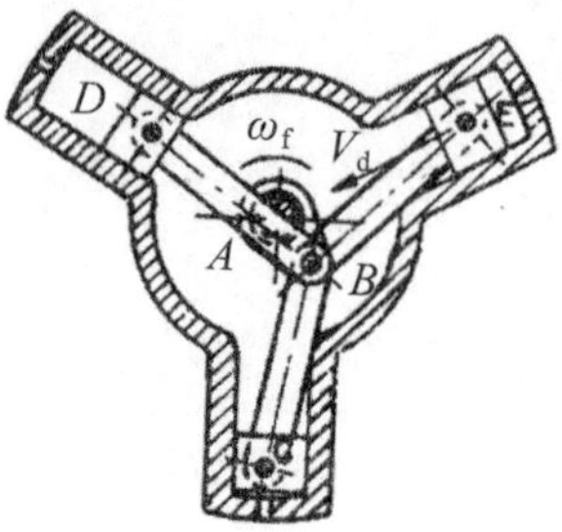

(c)

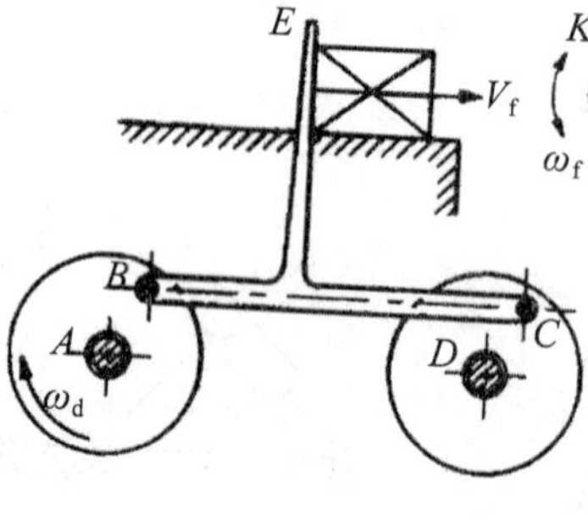

(d)

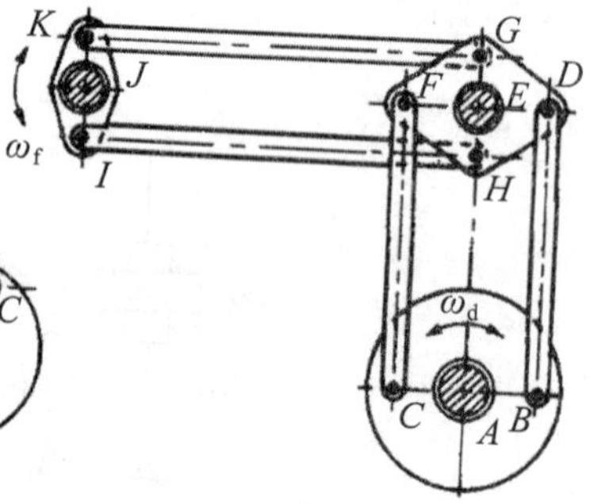

(e)

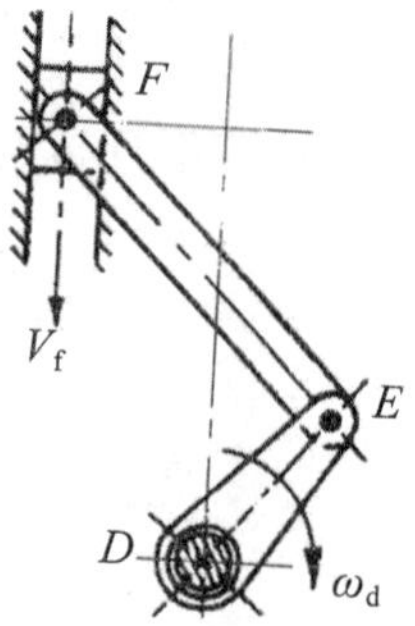

(f)

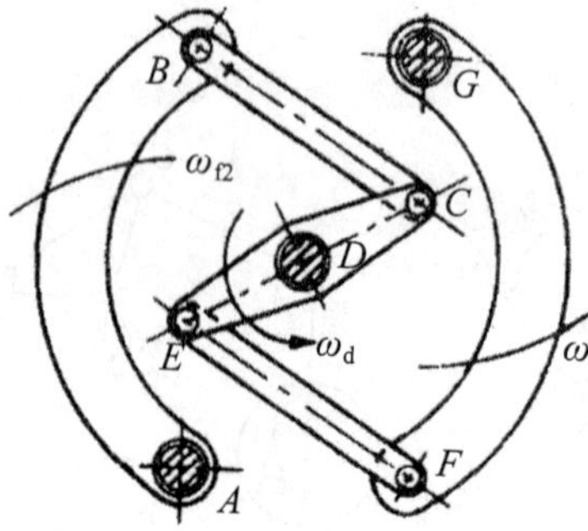

(g)

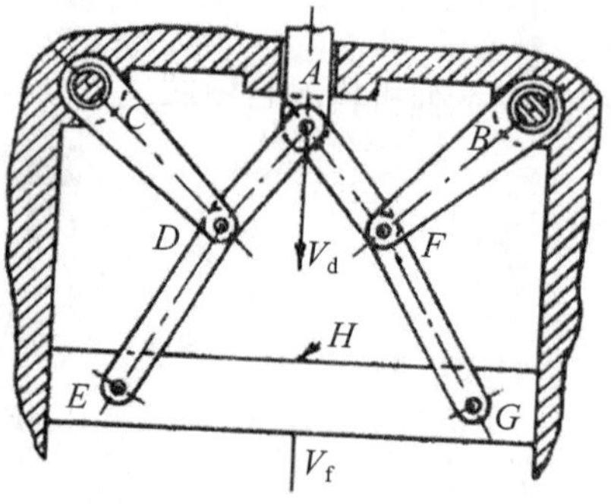

(h)

3.4 Open Question

Take a mechanism from life, draw a kinematical diagrams, calculate degrees of freedom, and analyze the relationship between degrees of freedom and the number of drivers.

7. 计算下列机构的自由度。

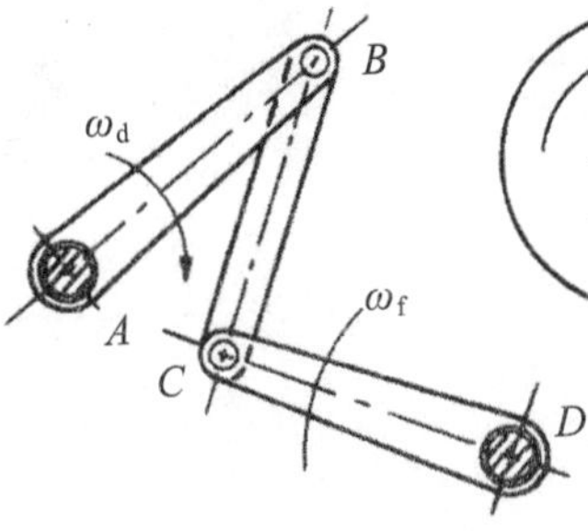

(a)

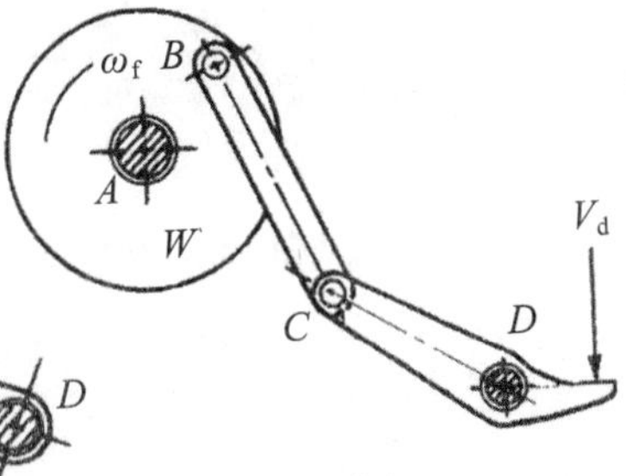

(b)

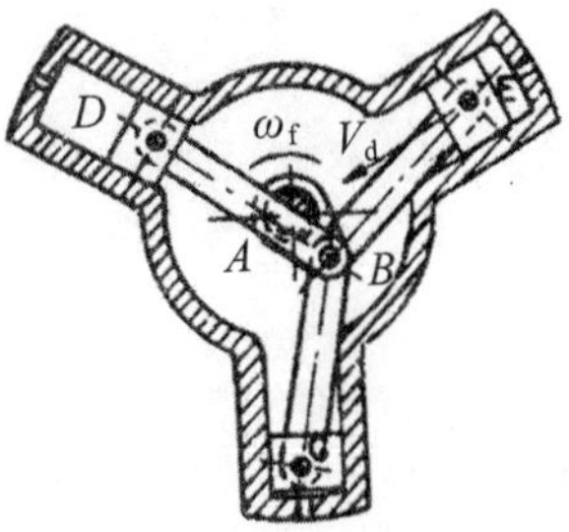

(c)

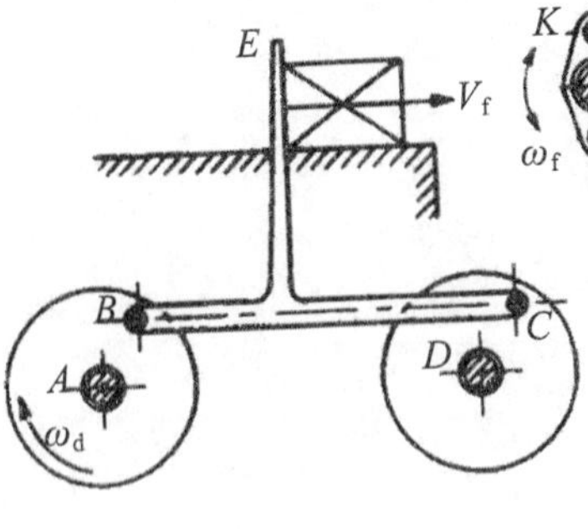

(d)

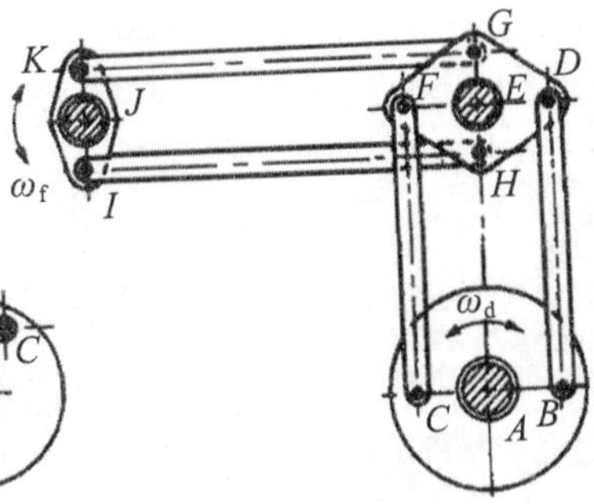

(e)

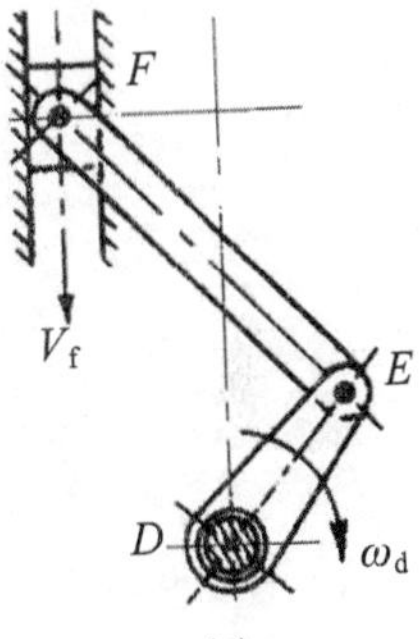

(f)

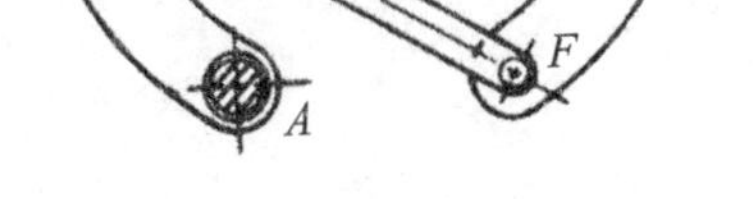

(g)

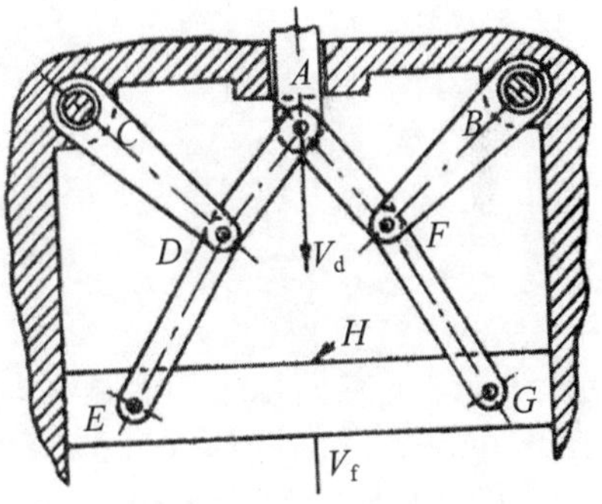

(h)

3.4 开放题

从生活中选取一个机构，画出结构简图，计算自由度，分析自由度和驱动数目的关系。

Chapter 4 Graphical Linkage Synthesis
第4章 连杆机构

4.1 Choice Questions

1. For a four-bar linkage system, the sum of the shortest and longest link of a planar quadrilateral linkage is less than or equal to the sum of the remaining two links, then in order to obtain the crank-rocker mechanism, the ground should take (　　).

A. the shortest link

B. the adjacent link of shortest link

C. the opposite link of shortest link

D. any link

2. For a four-link mechanism, the sum of the shortest and longest link of a planar quadrilateral linkage is less than or equal to the sum of the remaining two links, then in order to obtain the double-crank mechanism, the ground should take (　　).

A. the shortest link

B. the adjacent link of shortest link

C. the opposite link of shortest link

D. any link

3. For a four-link mechanism, the sum of the shortest and longest link of a planar quadrilateral linkage is less than or equal to the sum of the remaining two links, then in order to obtain the double-rocker mechanism, the ground should take (　　).

A. the shortest link

B. the adjacent link of shortest link

C. the opposite link of shortest link

D. any link

4. For a four-link mechanism, the sum of the shortest and longest link of a planar quadrilateral linkage is greater than or equal to the sum of the remaining two links, (　　).

A. the mechanism has a crank

B. the mechanism does not have a crank

C. sometimes the mechanism has a crank, sometimes it doesn't

D. None of the above answers are correct

5. The locked point of the mechanism is the position when the pressure angle α= (　　).

A. 0°　　　　B. 45°

C. 90°　　　　D. 135°

4.1 选择题

1. 铰链四杆机构中，若最短杆与最长杆长度之和小于或等于其余两杆长度之和，则为了获得曲柄摇杆机构，其机架应取（　　）。

A. 最短杆
B. 最短杆的相邻杆
C. 最短杆的相对杆
D. 任意一杆

2. 铰链四杆机构中，若最短杆与最长杆长度之和小于或等于其余两杆长度之和，则为了获得双曲柄机构，其机架应取（　　）。

A. 最短杆
B. 最短杆的相邻杆
C. 最短杆的相对杆
D. 任意一杆

3. 铰链四杆机构中，若最短杆与最长杆长度之和小于或等于其余两杆长度之和，则为了获得双摇杆机构，其机架应取（　　）。

A. 最短杆
B. 最短杆的相邻杆
C. 最短杆的相对杆
D. 任意一杆

4. 绞链四杆机构中，若最短杆与最长杆的长度之和大于其余两杆的长度之和，则（　　）。

A. 机构有曲柄存在
B. 机构不存在曲柄
C. 机构有时有曲柄，有时没有曲柄
D. 以上答案均不对

5. 机构的死点位置是指机构处于压力角α=（　　）时的位置。

A. 0°　　B. 45°
C. 90°　　D. 135°

6. When the crank is the driver, crank-rocker mechanism has________ locked point position, and when the rocker is the driver, the mechanism has________ locked point position. (   ).

A. a; a   B. a; no   C. no; a   D. no; no

7. When the angle of the extreme positions of the crank (   ), the crank-rocker mechanism has a quick return.

A. $\theta<0°$ B. $\theta=0°$ C. $\theta\leqslant0°$ D. $\theta>0°$

4.2 Fill-in-the-Blanks Questions

1. The locked point position of the four-link mechanism occurs when the follower and the coupler are ________ .

2. For the crank-rocker mechanism, when the crank rotates at the uniform speed, the average speed of the rocker reciprocating swing motion (back and forth) is different, which is called "________".

3. In the mechanism shown in the figure, when AB is the driver, the transmission angle of the mechanism is ________ .

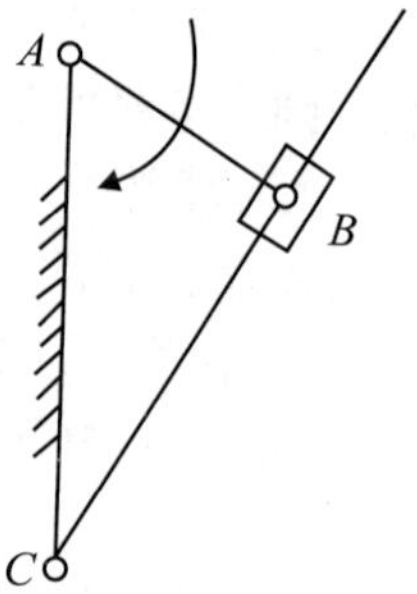

4. For the crank shaper mechanism, If the swing angle of the rocker ψ=30°, the speed ratio factor K is________.

Crank shaper mechanism

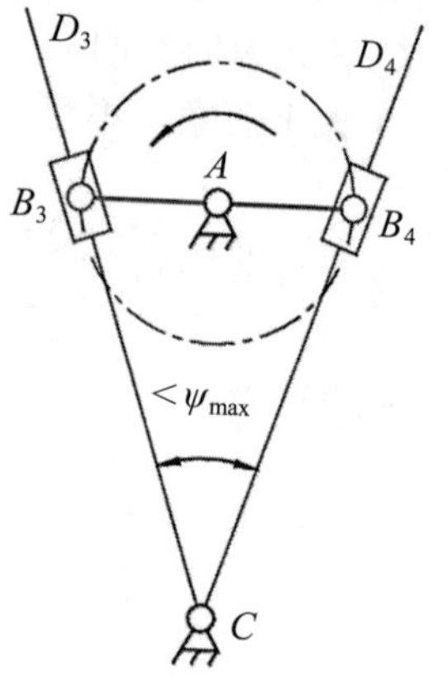

6. 曲柄摇杆机构，当曲柄为原动件时，机构________死点位置，而当摇杆为原动件时，机构________死点位置。（　　）

A. 有　有　　B. 有　无　　C. 无　有　　D. 无　无

7. 当曲柄的极位夹角为（　　）时，曲柄摇杆机构才有急回运动。

A. $\theta<0°$　　B. $\theta=0°$

C. $\theta\leq0°$　　D. $\theta>0°$

4.2 填空题

1. 铰链四杆机构的死点位置发生在从动件与连杆________位置。

2. 在曲柄摇杆机构中，当曲柄等速转动时，摇杆往复摆动的平均速度不同的运动特性称为________。

3. 在图示导杆机构中，*AB*为原动件时，该机构传动角的值为________。

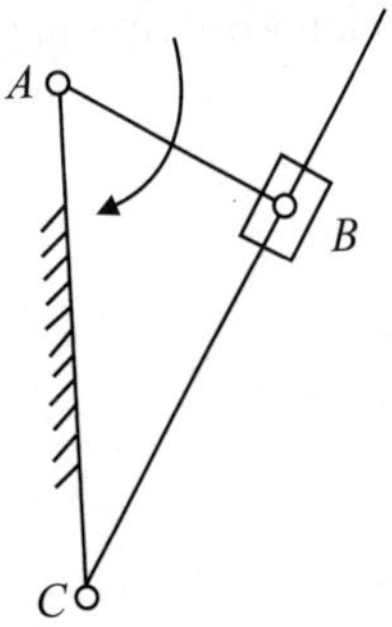

4. 在摆动导杆机构中，摇杆摆角ψ=30°，则其行程速度变化系数*K*的值为________。

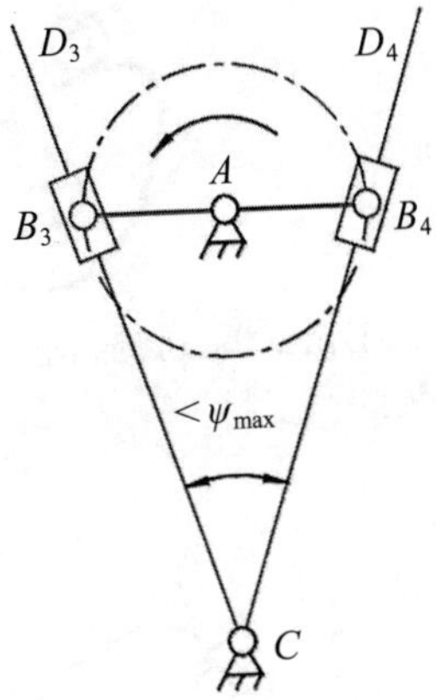

5. The crank-slider mechanism evolved from__________.

6. The planar linkage mechanism consists of some rigid members connected by __________and__________pairs.

7. The position of the mechanism when the pressure angle α=90° is called the __________of the mechanism.

8. When the crank is the driver, crank rocker mechanism has__________ locked point position, and when the rocker is the driver, the mechanism has__________ locked point position.

4.3 Analysis and Calculation Questions

1. The figure shows a hinged four-link mechanism with known link lengths: L_{AB}=10 cm, L_{BC}=25 cm, L_{CD}=20 cm, L_{AD}=30 cm. When the fixed links *AB*, *BC*, *CD* and *AD* are ground, which kind of mechanism do they belong to?

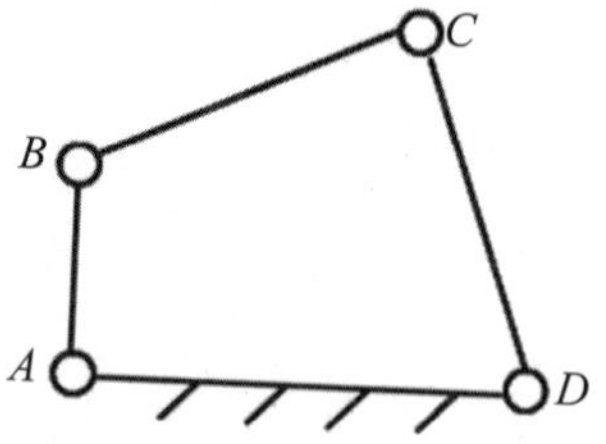

2. Draw the pressure angle of the following mechanism.

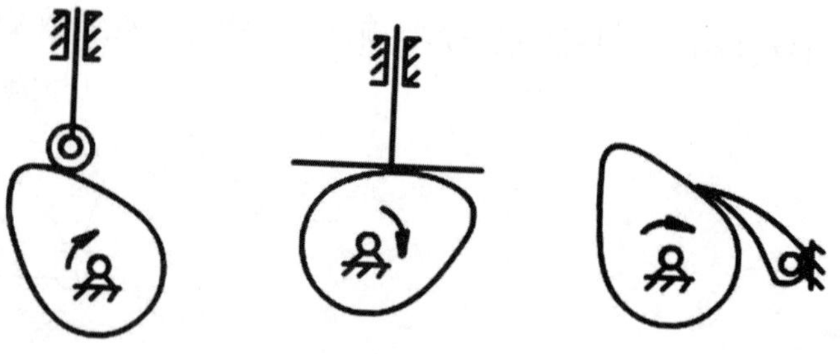

3. Draw the pressure angle of the following mechanism.

5. 曲柄滑块机构是由________机构演化而来的。

6. 平面连杆机构由一些刚性构件用________副和 ________副相互连接而组成。

7. 机构处于压力角α=90°时的位置，称为机构的________位置。

8. 曲柄摇杆机构，当曲柄为原动件时，机构________死点位置，而当摇杆为原动件时，机构________死点位置。

4.3 分析计算题

1. 图示为一铰链四杆机构，已知各杆长度：L_{AB}=10 cm，L_{BC}=25 cm，L_{CD}=20 cm，L_{AD}=30 cm。当分别固定构件*AB*、*BC*、*CD*、*AD*为机架时，它们各属于哪一类机构？

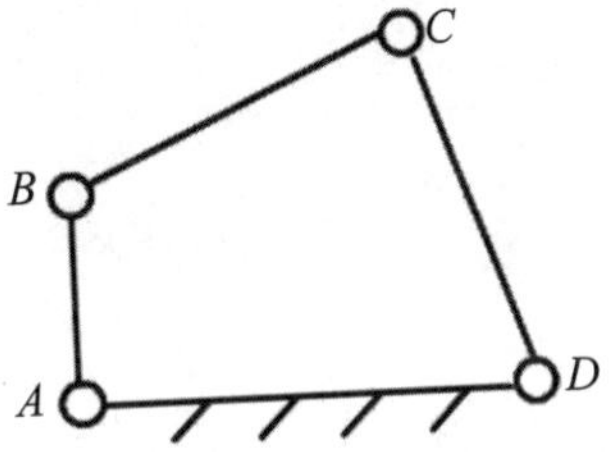

2. 绘制下列机构图示位置的压力角。

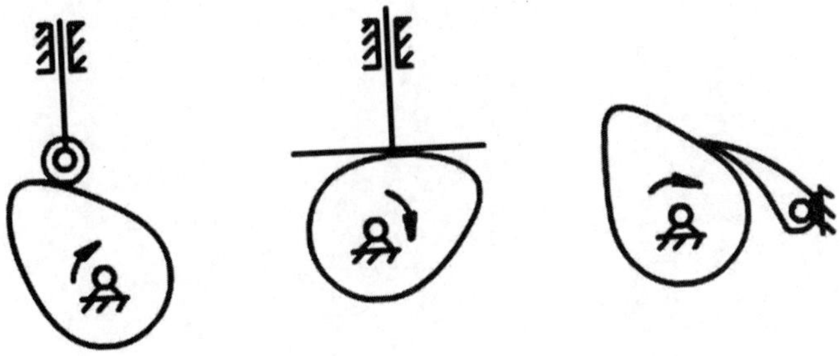

3. 绘制下列机构图示位置的压力角。

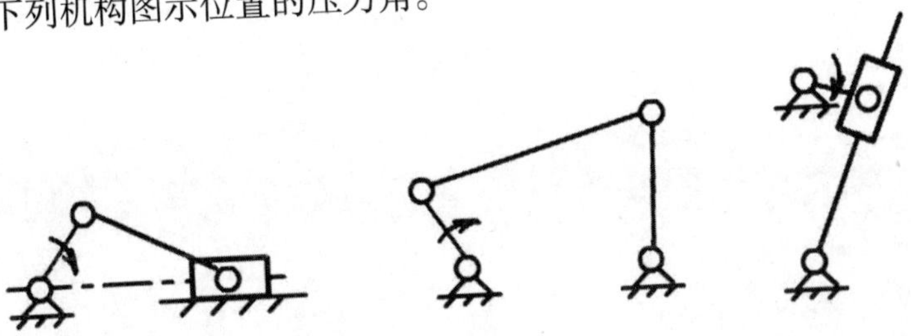

4. The picture show a door of heating furnace. The distance between the two movable joints on the door is 50 mm. When the furnace door opens, the temperature lower side is required to face up. Fixed joints are installed on the *y*-*y* line. The relevant dimensions are shown in the figure. Find the length of the other three links.

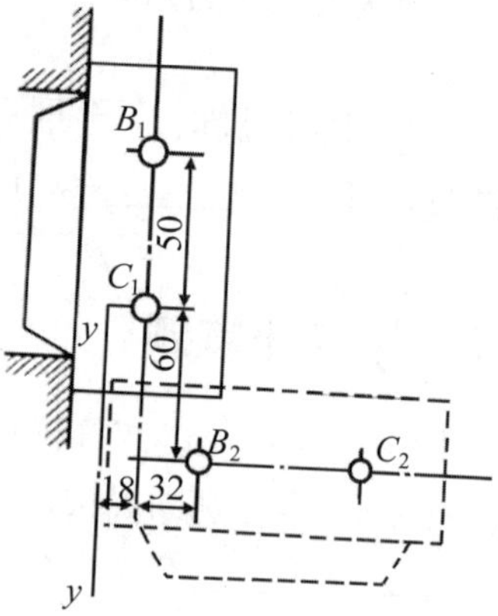

4. 如图所示为加热炉门。门上两个铰链距离为50 mm，当炉门打开时，要求温度较低的一侧朝上，固定铰链安装在y—y线上，相关尺寸如图所示，确定其他三个构件长度。

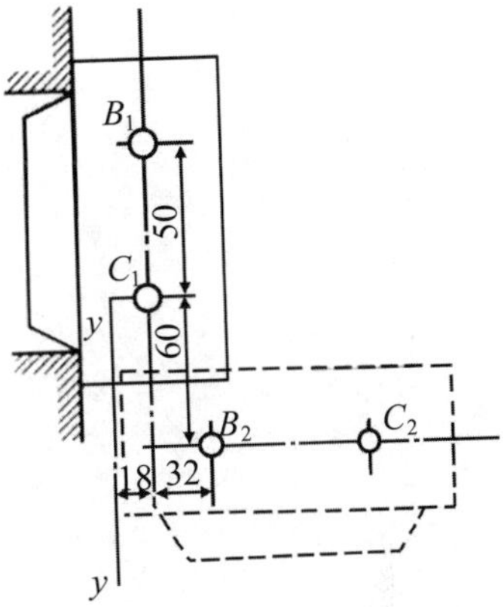

Chapter 5 Failure and Prevention
第5章 失效与预防

5.1 Choice Questions

1. The main failure form of rolling bearing at normal speed is (　　).
A. excessive plastic deformation　　B. excessive wear
C. fatigue pitting　　D. glue together

2. What type of stress is the variable load produced by? (　　).
A. Variable stress　　B. Static load stressing
C. Variable stress or static stress　　D. Variable stress and static stress

3. What type of stress is the static load produced by? (　　).
A. Variable stress　　B. Static load stressing
C. Variable stress or static stress　　D. Variable stress and static stress

4. Which of the following is not a load type? (　　).
A. Tension　　B. Bending
C. Torsion　　D. Impact

5. Which of the following is not a stage in the bathtub curve? (　　).
A. Infant-mortality stage　　B. Normal-life stage
C. Wear-out stage　　D. Complex failure stage

5.2 Fill-in-the-Blanks Questions

1. Variable stress can be generated by________load. Considering the conditions of the movement of the object, the static stress can be generated by the type of ________ load.

2. The strength criterion in the design criterion is ________________.

3. What stage do Ⅰ, Ⅱ and Ⅲ represent in the bathtub curve?
Ⅰ: ________; Ⅱ: ________; Ⅲ: ________ .

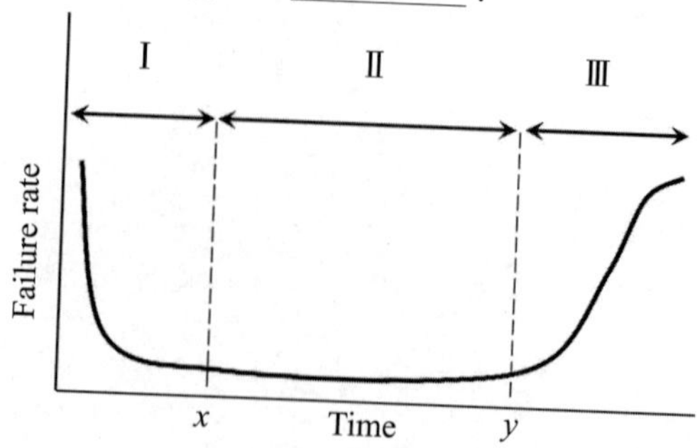

5.1 选择题

1. 滚动轴承在一般转速下的主要失效形式是（　　）。

A. 过大的塑性变形　　B. 过度磨损

C. 疲劳点蚀　　D. 胶合

2. 动载荷产生的应力是（　　）。

A. 动应力　　B. 静应力

C. 动应力或静应力　　D. 动应力和静应力

3. 静载荷产生的应力是（　　）。

A. 静应力　　B. 动应力

C. 静应力或动应力　　D. 静应力和动应力

4. 下列各项中不属于载荷类型的是(　　)。

A. 拉压　　B. 弯曲

C. 扭转　　D. 撞击

5. 下列哪一项不是“浴盆曲线”中的阶段？（　　）

A. 早期失效阶段　　B. 正常使用阶段

C. 磨损失效阶段　　D. 复合失效阶段

5.2 填空题

1. 动载荷能够产生的应力类型有________；考虑物体运动条件下，静载荷能够产生的应力类型有________。

2. 设计准则中的强度准则是______________。

3. 图示“浴盆曲线”中Ⅰ、Ⅱ、Ⅲ分别表示什么阶段？

Ⅰ: ________ ；Ⅱ: ________ ；Ⅲ: ________ 。

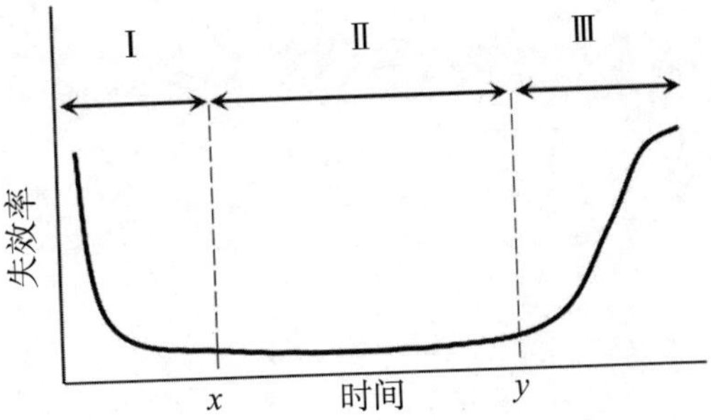

4. Two main characteristics of fatigue failure are____________ and____________.

5.3 Discussion Topics

1. Briefly explain the bathtub curve with an example.

2. Please review the four strength theories in material mechanics, and explain the formulas and applications of the four strength theories.

4. 疲劳失效的两个主要特征是__________和__________。

5.3 讨论题

1. 用一个例子简要解释“浴盆曲线”。

2. 请温习材料力学中的四大强度理论，并阐述四大强度理论的公式和应用场合。

Chapter 6 Shaft

第6章 轴

6.1 Choice Questions

1. What kinds of shaft are classified by shape? (　　)
A. Straight shaft　　B. Stepped shaft
C. Crankshaft　　D. Flexible wire shaft

2. According to the load carry and the role they play, shafts can be divided into (　　).
A. mandrel　　B. propeller shaft
C. rotating shaft　　D. stepped shaft

3. Which side is the working face of the woodruff key and the flat key? (　　)
A. Top side　　B. Side
C. Bottom side　　D. Neither

4. A good practice is to start with an inexpensive, (　　)for the first time through the design calculations.
A. low carbon steel　　B. medium carbon steel
C. high carbon steel　　D. low or medium carbon steel

5. Two or more keyways on the shaft should be (　　).
A. on the same generatrix
B. not on the same generatrix
C. Either A or B is acceptable

6. The main function of both couplings and clutches is to (　　).
A. compensate for the combined displacement of the two shafts
B. connect the two shafts to rotate and transmit torque
C. prevent overloading of the machine
D. mitigate shock and vibration

7. The shaft shoulder height should be (　　) than/to the inner ring height of the rolling bearing in order to disassemble the bearing.
A. greater　　B. smaller　　C. equal

8. The inner ring and shaft of rolling bearings are often fixed by a/an (　　) to achieve circumferential fixation.
A. key join　　B. press fit
C. spline join　　D. pin join

6.1 选择题

1. 轴按形状分类,可分为哪几种?( )

A. 直轴 B. 阶梯轴

C. 曲轴 D. 挠性钢丝轴

2. 轴按承受的载荷分类,可分为哪几种?( )

A. 心轴 B. 传动轴

C. 转轴 D. 阶梯轴

3. 半圆键、平键工作以( )为工作面。

A. 顶面 B. 侧面

C. 底面 D. 都不是

4. 在最初进行轴的设计计算时,材料一般选择( )。

A. 低碳钢 B. 中碳钢

C. 高碳钢 D. 低碳钢或中碳钢

5. 轴上两个以上的键槽( )。

A. 需要在同一个母线上

B. 不需要在同一个母线上

C. A项或B项都可以

6. 联轴器和离合器均具有的主要作用是( )。

A. 补偿两轴的综合位移

B. 连接两轴,使其旋转并传递转矩

C. 防止机器过载

D. 缓和冲击和振动

7. 为了方便拆卸轴承,轴肩高度应该( )轴承内圈外径。

A. 大于 B. 小于 C. 等于

8. 为了达到圆周上定位,轴承内圈和轴之间是( )。

A. 键连接 B. 压力配合

C. 花键连接 D. 销连接

9. The axial fixing of the outer ring of rolling bearings is often based on (　　).

A. shaft shoulder　　B. bearing end cover

C. round nut　　D. screw

6.2 Fill-in-the-Blanks Questions

1. Straight shafts are widely used and can be divided into smooth shaft and ________according to different shapes.

2. In order to facilitate the installation and disassembly parts, the stepped shaft is often designed in the shape of a__________middle and a__________end to facilitate the installation and disassembly parts from both ends.

3. The shaft segment that matches the bearing is called the__________.

4. The shaft segment on which the shaft hub (such as gears, pulleys) is installed on the shaft, called the__________.

5. The shaft that bears both bending moment and transmission torque when working is called the__________; The shaft that bears only bending moments and does not transmit torque is called the__________; The shaft used to transmit torque during operation without bearing the bending moment is called the __________.

6. The axial fixing methods of the parts on the shaft mainly include__________, round nut, shaft end retaining ring, etc.

6.3 Analysis and Calculation Questions

1. An air compressor consists of four compressor disks mounted on a steel shaft 50 mm apart, as shown in the following picture. The shaft is simply supported at either end. Each compressor wheel mass is 18 kg. Calculate the deflection at each wheel using Macaulay's method then calculate, using the Rayleighe-Ritz equation, the first critical frequency of the shaft. The shaft outer diameter is 0.05 m, the inner diameter is 0.03 m. (The calculation of critical frequency is an extension and not the required content of this chapter)

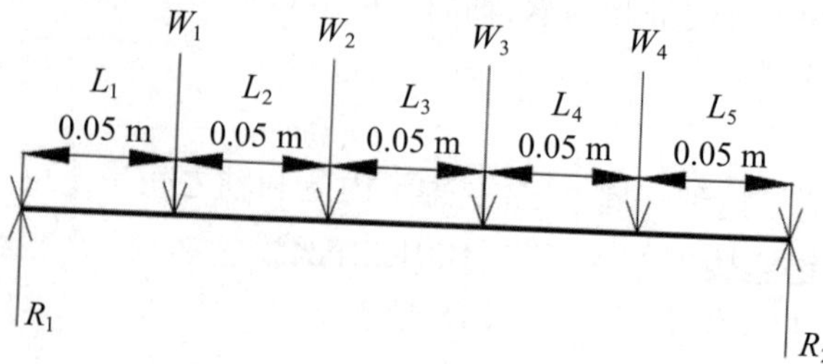

9. 轴承外圈是通过（　　）进行轴向定位。

A. 轴肩　　B. 轴承盖

C. 螺母　　D. 螺钉

6.2 填空题

1. 直轴应用较广泛，根据外形不同，可分为光轴和________。

2. 为了便于零件的装卸，阶梯轴常设计成中间________、两端________的形状，以便于零件从两端装卸。

3. 与轴承相配合的轴段，称为________。

4. 轴上安装轴毂（如齿轮、带轮）的轴段，称为________。

5. 工作时同时承受弯矩和传递扭距的轴，称为________；只受弯矩，不传递扭距的轴，称为________；工作时用于传递转矩而不承受弯矩的轴，称为________。

6. 轴上零件的轴向固定方法主要有________、圆螺母、轴端挡圈等。

6.3 分析计算题

1. 空气压缩机由4个安装在间隔50 mm的钢轴上的压缩盘组成，如下图所示。轴的两端都有简单的支撑。每个压缩机叶轮质量为18 kg。用麦考利法计算每个车轮的挠度，然后用瑞利-里兹公式计算轴的第一临界频率。轴的外径为0.05 m，内径为0.03 m。(临界频率的计算是拓展内容，不是本章要求内容)

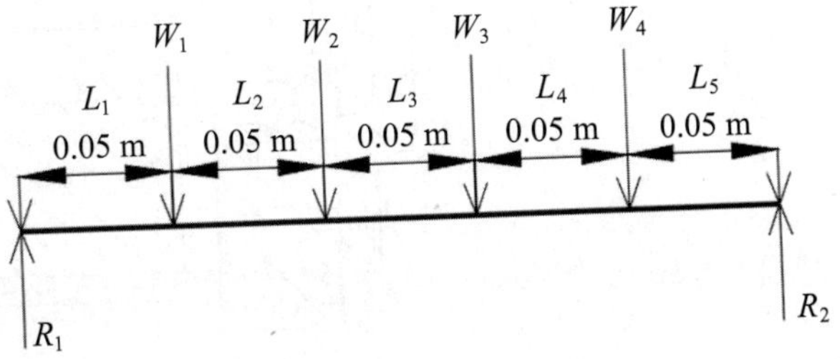

2. What is unreasonable design about the structure 1, 2 and 3 of the shaft shown in the diagram?

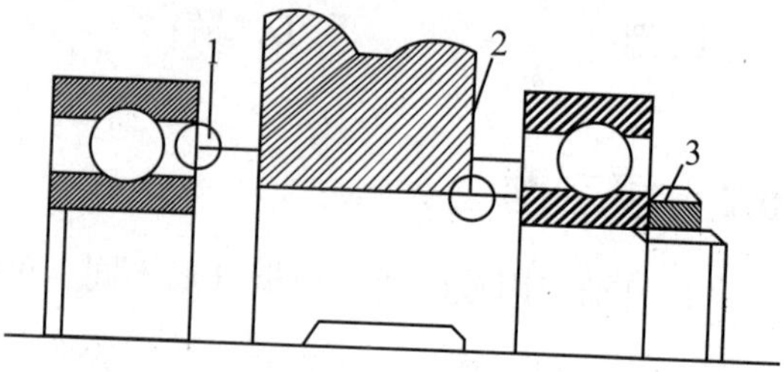

3. The figure shows the shafting structure with a pair of tapered roller bearings with narrow outer ring mounted against each other(face to face). Try to point out other structural errors (no less than 7) in the figure as shown in example ①.(Note: Lubrication method, chamfer and rounded corners are ignored.)

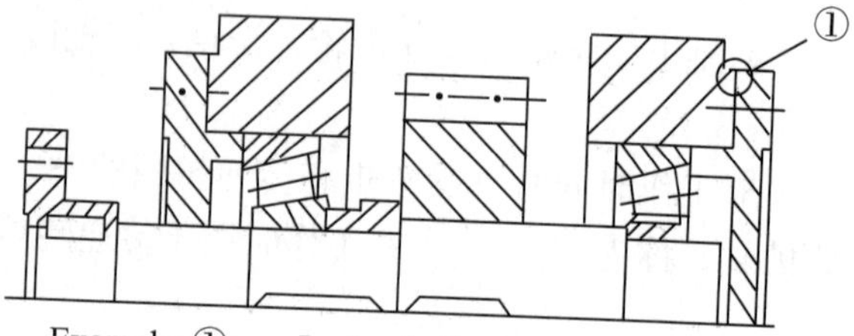

Example: ①——Lack of adjustment shim.

4. For shown the shafting structure diagram, try to analyze the 6 main errors.

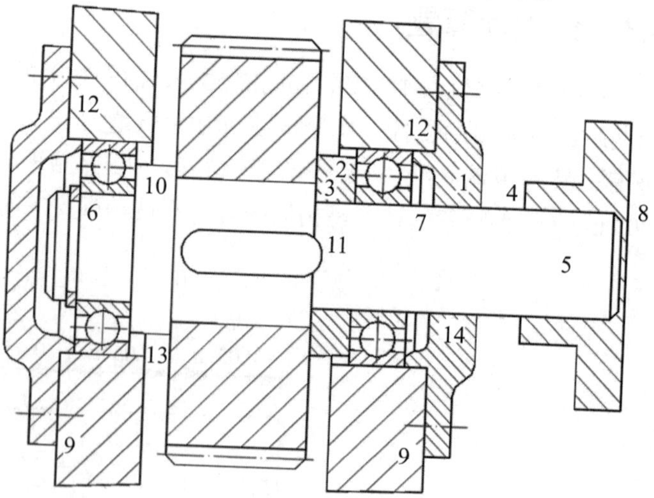

2. 图示轴的结构1、2、3处有哪些不合理的地方？用文字说明。

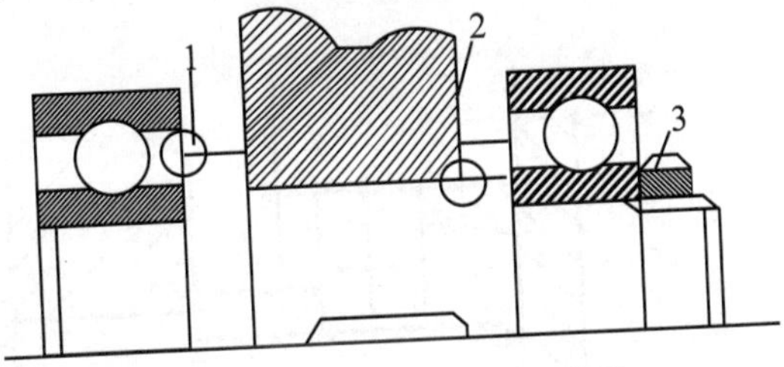

3. 图示为用一对圆锥滚子轴承外圈窄边相对安装的轴系结构。试按例①所示，指出图中的其他结构错误(不少于7处)。(注:润滑方式、倒角和圆角忽略不计。)

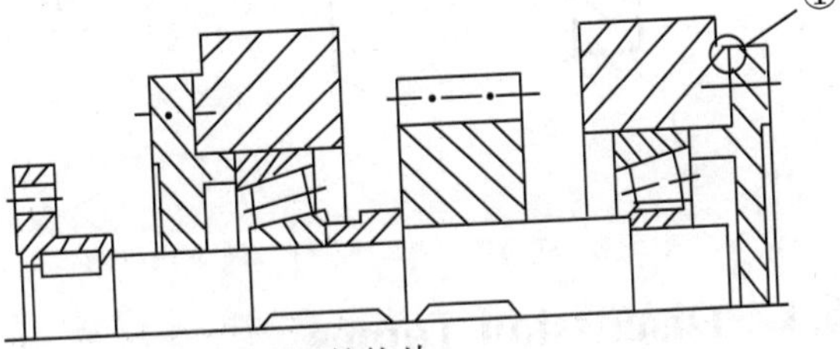

例:①——缺少调整垫片。

4. 如图所示轴系结构图，试分析其中的6处主要错误。

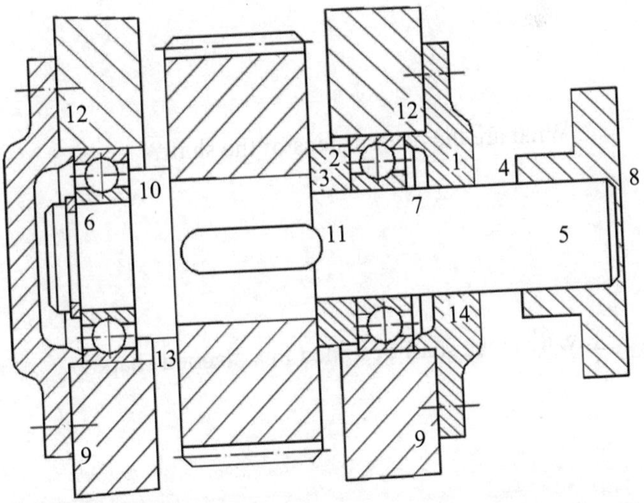

5. Please find the unreasonable designs on the shaft layout and explain the reasons.

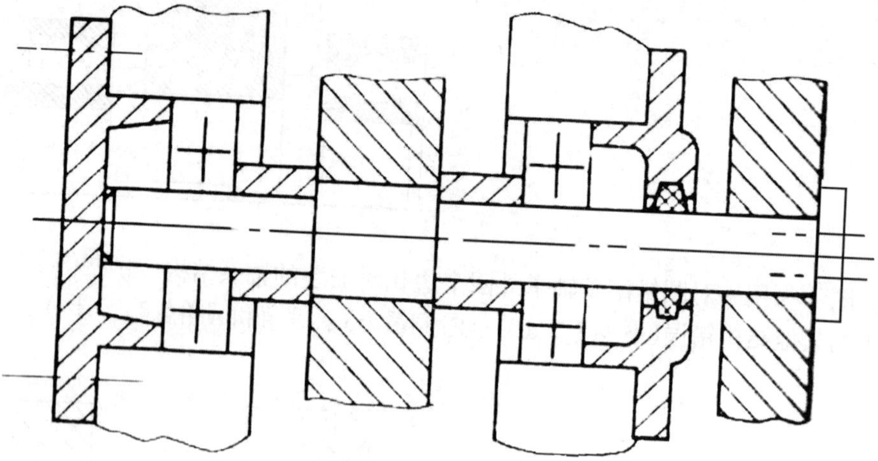

6.4 Discussion Topics

1. What kind of shaft do the front and rear shaft of a bicycle belong to according to the load bearing classification?

2. What are the design steps of the shaft?

3. Why is the shaft designed in a stepped shape?

5. 请指出并说明图示轴系结构中的错误。其中,轴承采用润滑油润滑,齿轮也采用油润滑,直角、圆角忽略不计。不同地方的相同错误算不同错误。

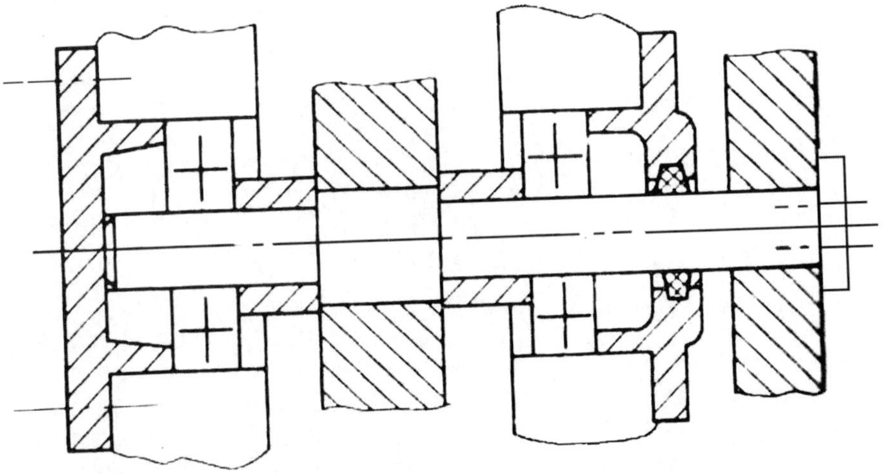

6.4 讨论题

1. 根据承受载荷分类,自行车的前轴和后轴各属于哪一种轴?

2. 轴的设计步骤是什么?

3. 为什么大多数的轴设计成阶梯形?

Chapter 7 Rolling-Contact Bearings
第7章　滚动轴承

7.1 Choice Questions

1. The main form of failure of rolling-contact bearings at general speeds is (　　).

A. excessive plastic deformation　　B. excessive wear

C. fatigue pitting　　D. gluing

2. Contact angle of deep groove bearing is(　　).

A. 20°　　B. 45°

C. 0°　　D. 60°

3. Contact angle of thrust bearing is (　　).

A. 0°　　B. 45°

C. 60°　　D. 90°

4. What type of bearing is the code 30310? (　　)

A. Angular contact ball bearing　　B. Cylindrical roller bearings

C. Needle roller bearings　　D. Tapered roller bearings

5. What type of bearing is the code 6207? (　　)

A. Angular contact ball bearings　　B. Deep groove bearings

C. Needle roller bearings　　D. Tapered roller bearings

6. Which of the following options is not an advantage of plain or slider bearings? (　　)

A. Small friction resistance and sensitive starting

B. Large load carrying capacity.

C. Impact resistance, stable work

D. Good high-speed performance

7. The rolling bearing component that evenly separate the rolling elements and avoid the rolling elements from contacting each other to reduce friction and wear is (　　).

A. cage(separator)　　B. outer ring

C. ball or roller　　D. inner ring

8. From economic considerations, as long as it can meet the requirements of use, which bearing should be preferred? (　　)

A. Cylindrical roller bearings　　B. Ball bearings

C. Spherical roller bearings　　D. Needle roller bearings

7.1 选择题

1. 滚动轴承在一般转速下的主要失效形式是（　　）。

A. 过大的塑性变形　　B. 过度磨损

C. 疲劳点蚀　　D. 胶合

2. 深沟球轴承的接触角是（　　）。

A. 20°　　B. 45°

C. 0°　　D. 60°

3. 推力轴承的接触角是（　　）。

A. 0°　　B. 45°

C. 60°　　D. 90°

4. 代码是30310的轴承类型是什么?（　　）

A. 角接触球轴承　　B. 圆柱滚子轴承

C. 滚针轴承　　D. 圆锥滚子轴承

5. 代码是6207的轴承类型是什么?（　　）

A. 角接触球轴承　　B. 深沟球轴承

C. 滚针轴承　　D. 圆锥滚子轴承

6. 下列各项中,不属于滑动轴承优点的是（　　）。

A. 摩擦阻力小、起动灵敏　　B. 承载能力大

C. 抗冲击,工作平稳　　D. 高速性能好

7. 把滚动体均匀隔开,避免滚动体相互接触,以减少摩擦与磨损的滚动轴承部件是（　　）。

A. 保持架　　B. 外圈

C. 滚动体　　D. 内圈

8. 从经济性考虑,只要能满足使用要求,应优先选（　　）轴承。

A. 圆柱滚子　　B. 球

C. 球面滚子　　D. 滚针

9. Rolling bearing 6120 indicates an inner diameter of (   ) mm.

A. 20
B. 100
C. 120
D. 12

10. The rating life of a group of nominally identical ball or roller bearings is defined as the number of revolutions (or hours at a constant speed) that (   )percent of a group of bearings will achieve or exceed before the failure criterion develops.

A. 50
B. 80
C. 90
D. 100

11. In (   ), balls are fitted well into the deep grooves, enabling the bearing to support axial loads in both directions.

A. deep groove bearings
B. ball bearings
C. spherical roller bearings
D. needle roller bearings

12. A (   ) can support an axial load in one direction. These bearings are designed not to accommodate radial loads. The components of these bearings can be easily separated.

A. deep groove bearings
B. ball bearings
C. thrust ball bearing
D. needle roller bearings

13.The ability of angular contact bearings to withstand axial loads (   )with increasing of the contact angle α.

A. increases
B. reduces
C. is unchanged
D. are not certain

7.2 Fill-in-the-Blanks Questions

1. The higher the viscosity of the lubricant selected in plain bearings, the ______ its load carrying capacity.

2. The fit between the inner ring and shaft adopts the______, and the fit between the outer ring and the housing adopts the________.

3. What are the coefficients X and Y in the calculation formula of the equivalent dynamic load P of rolling bearings? X: ________.Y: ______.

9. 滚动轴承6120表示内径为（　　）mm。

A. 20　　B. 100

C. 120　　D. 12

10.一组同一型号轴承在同一条件下运转，其可靠度为（　　）时，能达到或超过的寿命称为基本额定寿命。

A. 50%　　B. 80 %

C. 90%　　D. 100%

11.（　　）中，滚珠与深沟槽配合良好，使轴承能够支撑两个方向的轴向载荷。

A. 深沟球轴承　　B. 球轴承

C. 球面滚子轴承　　D. 滚针轴承

12.（　　）可以在一个方向上承受轴向载荷，这种轴承的设计不适应径向载荷，其部件可以很容易地分离。

A. 深沟球抽承　　B. 球轴承

C. 推力球轴承　　D. 滚针轴承

13.角接触轴承承受轴向载荷的能力，随接触角α的增大而（　　）。

A. 增大　　B. 减少

C. 不变　　D. 不定

7.2 填空题

1. 滑动轴承中所选用的润滑油，其黏度越大，则其承载能力越________。

2. 滚动轴承内圈与轴颈的配合采用____________制，外圈与轴承孔的配合采用____________制。

3. 滚动轴承的当量动载荷P的计算公式中，X和Y是什么系数?

X:__________；Y:__________。

4. ______have long and thin rollers, these bearings are used for applications where radial space is limited.

5. ______are able to withstand a large thrust load in a single direction, in addition to radial loads.

6. The basic rating life of a bearing refers to the total number of revolutions before ______(percentage) of bearings of the same type operate without fatigue pitting when the same type of bearing is operating under the same conditions.

7. The seeds of failure for bearings are : ________, inclusions, impact, cyclic high stress, degradation of the lubricant.

8. Life index ε (α in English textbook) is______for ball bearings and________for roller bearsing (cylindrical and tapered roller).

7.3 Analysis and Calculation Questions

1.A mechanical shaft is supported at both ends by 6313 deep groove ball bearings. Each bearing bears a radial load F_r=5400 N and an axial load F_a=2650 N, with slight impact. The shaft speed is n=1250 r/min, and the expected service life L_h=5000 h. Is this bearing suitable?

2. Take SKF bearing as example, which rates its bearings for 1 million revolutions, if you desire a life of 5000 h at 1725 rev/min with a load of 400 lbf (1 lbf=4.45 N)with a reliability of 90%, calculate the catalog load rating.

4. ________具有细长滚子，这些轴承用于径向空间有限的应用。

5. 除了径向载荷，_______还能够承受单个方向上的大的轴向载荷。

6. 轴承的基本额定寿命是指同型号的轴承在相同条件下运转时，_______（占比）的轴承未发生疲劳点蚀前运转的总转数。

7. 轴承发生疲劳失效的原因有：________、杂质、冲击、高循环应力、润滑不够。

8. 寿命指数 ε（英文教材中用α）对于球轴承是_______，对于滚子轴承是_______。

7.3 分析计算题

1. 一机械转轴，两端用6313深沟球轴承支撑，每个轴承承受的径向载荷F_r=5 400 N，轴向载荷为F_a=2 650 N，温度低于100℃，运行时有轻微冲击，轴的转速为n = 1 250 r/min，预期寿命L_h=5 000 h，此轴承是否合适？

2. 以SKF轴承为例，其轴承转速为100万转。如果希望购买到在1725 r/min的转速下，负载400 lbf（1lbf = 4.45 N），可靠性为90%，使用5000 h的轴承，计算其基本额定动载荷。

3. As shown in the figure, the active shaft in a certain machinery is intended to be supported by a pair of angular contact ball bearings, please judge whether it is suitable. The primary bearing designation is 7211AC. It is known that the speed of the shaft n = 1 450 r/min, the radial load of the two bearings is F_{r1}=3 300 N, F_{r2}=1 000 N, axial load F_A=900 N, the bearing works at room temperature, there is a medium impact during operation, C_r=50 500 N.

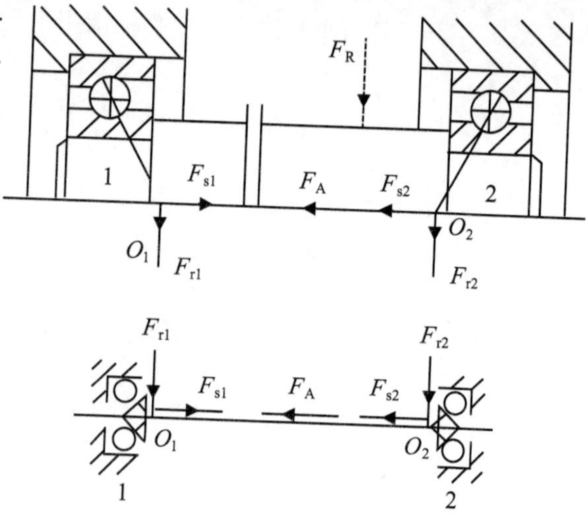

Tapered roller bearing	Angular contact ball bearings		
	70 000C(α=15°)	70 000AC(α=25°)	70 000B(α=40°)
$F_S/F_r/(2Y)$	F_S=e F_r	F_S=0.68 F_r	F_S=1.14 F_r

Type of bearing		F_a/C_o	e	$F_a/F_r>e$		$F_a/F_r\leqslant e$	
				X	Y	X	Y
Angular contact ball bearings	α=25°	—	0.68	0.41	0.87	1	0
	α=40°	—	1.14	0.35	0.57	1	0

Coefficient of loading f_p

Load properties	No impact or slight impact	Medium impact	Strong impact
f_p	1.0～1.2	1.2～1.8	1.8～3.0

7.4 Discussion Topics

1. What are the characteristics of needle bearings?

2. What is the difference between basic rated dynamic load C (catalog load rating) and equivalent dynamic load P?

3. 如图所示为某机械中的主动轴，拟用一对角接触球轴承支承，请判断是否合理。初选轴承型号为 7211AC。已知轴的转速 n=1 450 r/min，两轴承所受的径向载荷分别为 F_{r1}=3 300 N，F_{r2}=1 000 N，轴向载荷 F_A=900 N，轴承在常温下工作，运转时有中等冲击，C_r=50 500 N。

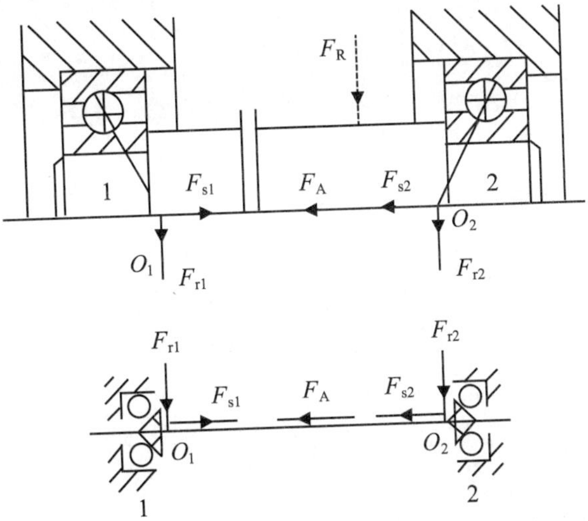

圆锥滚子轴承	角接触球轴承		
	70 000C(α=15°)	70 000AC(α=25°)	70 000B(α=40°)
$F_s/F_r/(2Y)$	$F_s=e\,F_r$	F_s=0.68 F_r	F_s=1.14 F_r

轴承类型		F_a/C_o	e	$F_a/F_r>e$		$F_a/F_r\leq e$	
				X	Y	X	Y
角接触球轴承	α=25°	—	0.68	0.41	0.87	1	0
	α=40°	—	1.14	0.35	0.57	1	0

载荷系数 f_p

载荷性质	无冲击或轻微冲击	中等冲击	强烈冲击
f_p	1.0～1.2	1.2～1.8	1.8～3.0

7.4 讨论题

1. 滚针轴承有什么特点?

2. 基本额定动载荷 C 与当量动载荷 P 有何区别?

Chapter 8 Gears 1
第8章 齿 轮 1

8.1 Choice Questions

1. Which one is not the type of gears? (　　)

A. Spur gear　　B. Involute gear

C. Bevel gear　　D. Helical gear

2. Pressure angle of base circle is (　　)

A. 0°　　B. 20°

C. 15°　　D. 25°

3. For full depth teeth, coefficient of addendum h_a^* and coefficient of clearance c^* are (　　).

A. $h_a^*=1.25$, $c^*=0.2$　　B. $h_a^*=1.25$, $c^*=0.3$

C. $h_a^*=1$, $c^*=0.25$　　D. $h_a^*=0.8$, $c^*=0.3$

4. The conditions of meshing of spur gears are (　　).

A. $m_1=\alpha_1$, $m_2=\alpha_2$　　B. $\alpha_1=\alpha_2$

C. $m_1=m_2$　　D. $m_1=m_2$, $\alpha_1=\alpha_2$

5. The contact ratio of spur gear is (　　).

A. $\varepsilon \geqslant 1$　　B. $\varepsilon = 1$

C. $\varepsilon < 1$　　D. $\varepsilon = 0$

6. The smallest number of teeth for spur pinion that will operate with a rack without interference is (　　).

A. 20　　B. 18

C. 17　　D. 13

7. The smallest number of teeth on a spur pinion and gear, 1 one-to-one gear ratio which can exist without interference is (　　).

A. 20　　B. 18

C. 17　　D. 13

8. The shape of the tooth profile of an involute gear depends on (　　).

A. pitch circle　　B. addendum circle

C. dedendum circle　　D. base circle

8.1 选择题

1. 下列各项中，不属于齿轮的基本类型的是（　　）。

A. 直齿轮　　B. 渐开线齿轮

C. 锥齿轮　　D. 斜齿轮

2. 基圆的压力角是（　　）。

A. 0°　　B. 20 °

C. 15°　　D. 25°

3. 对于标准齿轮, 齿顶高系数 h_a^* 和顶隙 c^* 分别是（　　）。

A. h_a^*=1.25，c^*=0.2　　B. h_a^*=1.25，c^*=0.3

C. h_a^*=1，c^*=0.25　　D. h_a^*=0.8，c^*=0.3

4. 直齿圆柱齿轮正确啮合的条件是（　　）。

A. $m_1=\alpha_1$, $m_2=\alpha_2$　　B. $\alpha_1=\alpha_2$

C. $m_1=m_2$　　D. $m_1=m_2$, $\alpha_1=\alpha_2$

5. 直齿圆柱齿轮的重合度应该是（　　）。

A. $\varepsilon \geqslant 1$　　B. $\varepsilon = 1$

C. $\varepsilon < 1$　　D. $\varepsilon = 0$

6. 与齿条一起啮合而没有干涉的直齿轮的最小齿数是（　　）。

A. 20　　B. 18

C. 17　　D. 13

7. 没有干涉的情况下一对1：1传动比的直齿轮啮合，其小齿轮最小齿数是（　　）。

A. 20　　B. 18

C. 17　　D. 13

8. 渐开线齿轮的齿廓曲线形状取决于（　　）。

A. 分度圆　　B. 齿顶圆

C. 齿根圆　　D. 基圆

9. What can be used to achieve transmission between two intersecting shafts? (　　)

A. Worm transnission

B. Helical cylindrical gear transmission

C. Straight bevel gear transmission

D. Straight tooth cylindrical gear transmission

10. What are the correct meshing conditions for ordinary cylindrical worm and worm gear transmission? (　　)

A. $m_{t1}=m_{a2}$, $\alpha_{t1}=\alpha_{a2}$, $\lambda=\beta$　　B. $m_{a1}=m_{t2}$, $\alpha_{a1}=\alpha_{t2}$, $\lambda=\beta$

C. $m_{t1}=m_{a2}$, $\alpha_{t1}=\alpha_{a2}$, $\lambda=-\beta$　　D. $m_{a1}=m_{a2}$, $\alpha_{a1}=\alpha_{t2}$, $\lambda=-\beta$

(Note: The subscript t represents the end face, a represents the axial direction, 1 represents the worm, and 2 represents the worm gear.)

11. The idler gear in the gear train will affect (　　).

A. gear train ratio　　B. gear speed

C. gear steering　　D. A,B and C

8.2 Fill-in-the-Blanks Questions

1. When the gears mesh, the____________ of the driving gear pushes the ____________ of the driven gear, a pair of teeth start to mesh, so the starting meshing point should be the intersection point of the____________ of the driven gear and the meshing line; when the ________ of the driving gear pushes at the____________ of the driven gear, the teeth of the two gears are about to be out of mesh, so the point of termination of meshing is the intersection of the____________ of the driving wheel and the meshing line.

2.The condition meshing of spur gears (involute standard gear) is that the ______ and ______ are equal respectively.

3.The commonly used processing methods for involute tooth profile include ______ and ______.

4.The correct meshing conditions for ordinary cylindrical worm and worm gear drives are____________.

5.When the gear train moves, all the geometric axes of the gears are fixed, which is called a fixed-axis gear train, and at least one gear whose geometric axis is not fixed is called an____________.

9. 要实现两根相交轴之间的传动，可以采用（　　）。

A. 蜗杆传动

B. 斜齿圆柱齿轮传动

C. 直齿锥齿轮传动

D. 直齿圆柱齿轮传动

10. 普通圆柱蜗杆和蜗轮传动的正确啮合条件是（　　）。

A. $m_{t1}=m_{a2}$，$\alpha_{t1}=\alpha_{a2}$，$\lambda=\beta$

B. $m_{a1}=m_{t2}$，$\alpha_{a1}=\alpha_{t2}$，$\lambda=\beta$

C. $m_{t1}=m_{a2}$，$\alpha_{t1}=\alpha_{a2}$，$\lambda=-\beta$

D. $m_{a1}=m_{a2}$，$\alpha_{a1}=\alpha_{t2}$，$\lambda=-\beta$

（注：下标 t 表示端面，a 表示轴向，1 表示蜗杆，2 表示蜗轮。）

11. 轮系中惰轮会影响（　　）。

A. 轮系传动比

B. 齿轮旋转速度

C. 齿轮旋转方向

D. A、B 、C 项都对

8.2 填空题

1. 齿轮啮合时，当主动齿轮的__________推动从动齿轮的__________，一对轮齿开始进入啮合，所以开始啮合点应为从动轮__________与啮合线的交点；当主动齿轮的__________推动从动齿轮的__________，两轮齿即将脱离啮合，所以终止啮合点为主动轮__________与啮合线的交点。

2. 渐开线标准直齿圆柱齿轮正确啮合的条件为________和________分别相等。

3. 渐开线齿形常用的加工方法有________和________两类。

4. 普通圆柱蜗杆和蜗轮传动的正确啮合条件是________。

5. 轮系运动时，所有齿轮几何轴线都固定不动的，称为定轴轮系，至少有一个齿轮几何轴线不固定的，称为________系。

6. The standard gear parameters are: ________, ________, ________, ________, and________.

8.3 Analysis and Calculation Questions

1. The worm transmission is shown in the figure. Knowing the helix direction and rotation direction of the worm, try to find the direction of rotation of the worm gear.

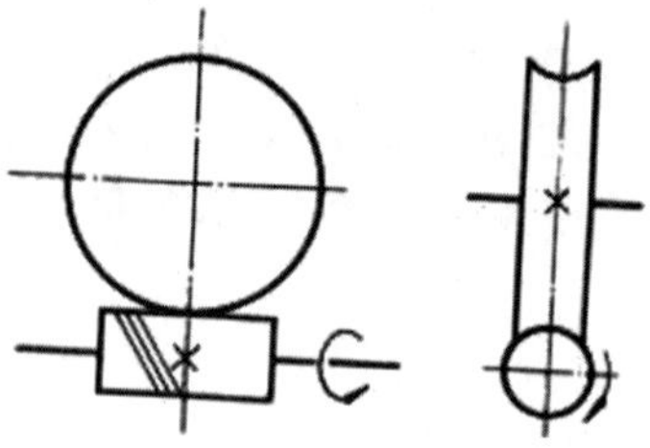

(a)

2. As shown in the figure, the rotation direction of the bevel gear Z_4 on the output shaft is known. In order to offset part of the axial force on the shaft Ⅱ, try to find: mark the rotation direction of each wheel on the diagram; determine the helix direction of the worm gear-sets (worm, worm gear), the direction of the forces on the worm and the worm gear, and the direction of the axial force on the bevel gear Z_3. (It is required to mark on the picture or draw another picture)

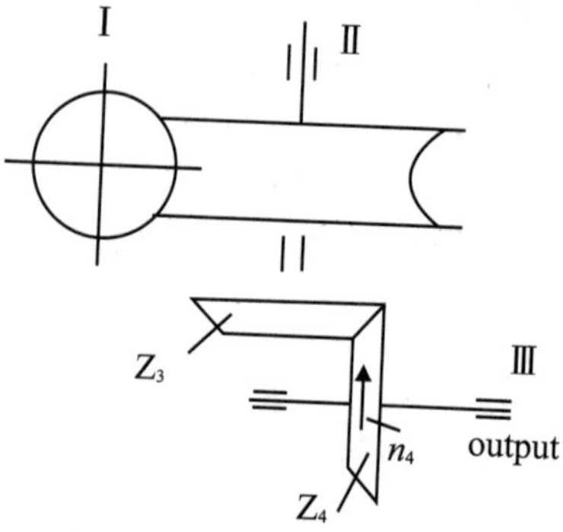

3. z_1=15, z_2=25, z_3=20, z_4=60. n_1 =200 r/min (clockwise), n_4=50 r/min (clockwise). Try to find the speed of H.

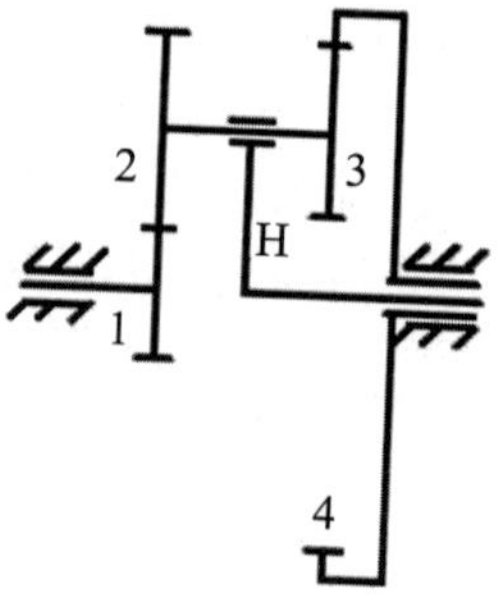

6. 标准渐开线直齿轮参数有________、________、________、________和________。

8.3 分析计算题

1. 如图所示蜗杆传动，已知蜗杆的螺旋线旋向和旋转方向，试求蜗轮的转向。

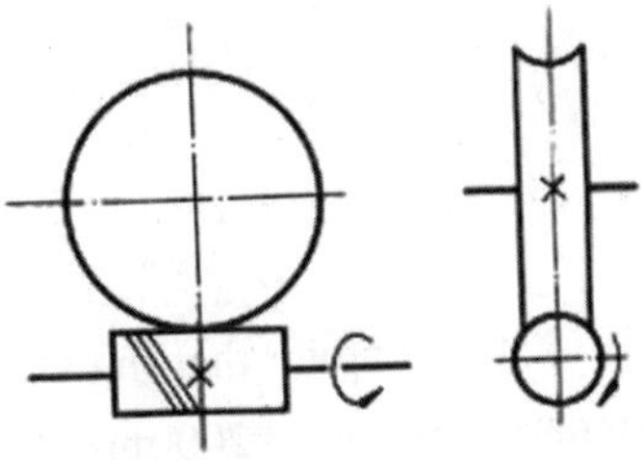

(a)

2. 如图所示，已知输出轴上的锥齿轮Z_4的转向n_4，为了使中间轴Ⅱ上的轴向力能被抵消一部分，试求：在图上标出各轮的转向；判断蜗杆传动的螺旋角方向（蜗杆、蜗轮），蜗杆、蜗轮所受各力方向以及锥齿轮Z_3所受轴向力方向。（要求标在图上或另画图表示）。

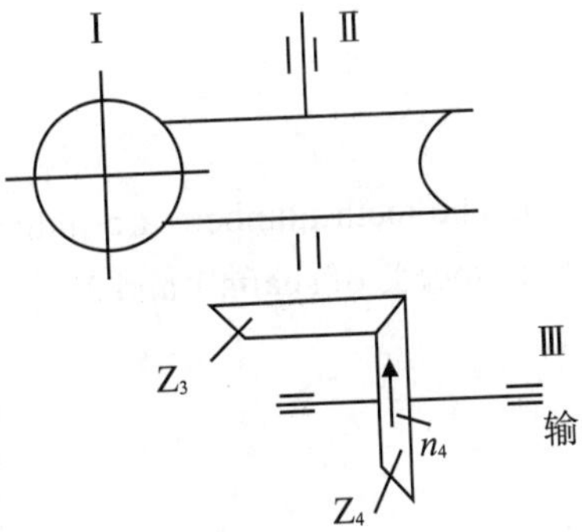

3. 轮系如图所示，齿数分别是z_1=15，z_2=25，z_3=20，z_4=60。n_1 =200 r/min（顺时针），n_4=50 r/min（顺时针）。试求H的转速。

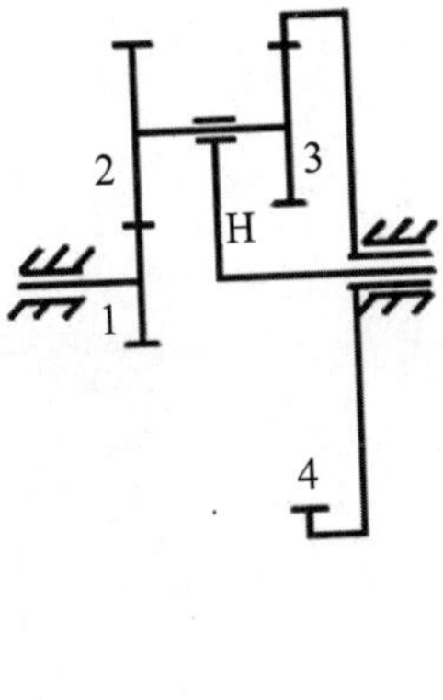

4. The teeth numbers of each gear in the gear train are given, z_1=20, z_2=18, z_3=56 respectively. Find the transmission ratio i_{1H}.

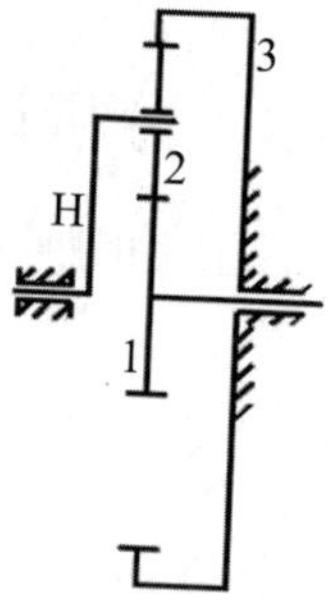

5. It is known that the standard center distance of a pair of external meshing involute standard gears is a=200 mm, the number of teeth z_1=20, z_2=80, the coefficient of addendum h_a^*=1, and the pressure angle α=20°. Find the modulus m, transmission ratio i_{12}, pitch circle diameter and meshing angle of the pair of gears.

6. The teeth numbers are indicated in the figure. The motor is driven at 1 750 r/min, find the speeds of shafts 1 and 2.

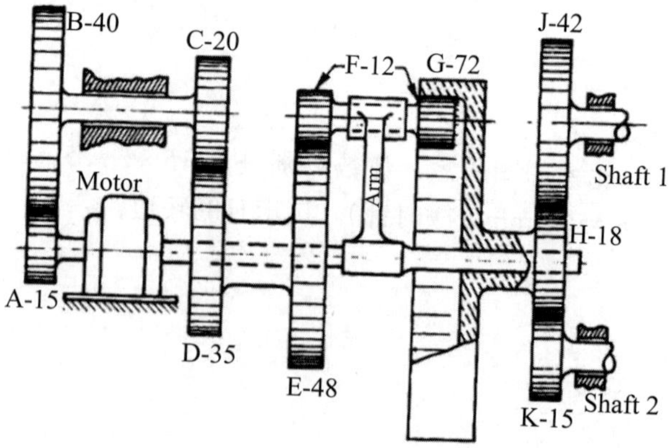

4. 已知轮系中各齿轮的齿数分别为z_1=20，z_2=18，z_3=56。求传动比i_{1H}。

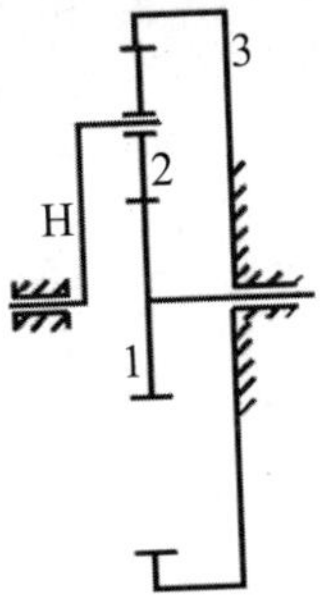

5. 已知一对外啮合渐开线标准直齿圆柱齿轮的标准中心距a=200mm，齿数z_1=20，z_2=80，齿顶高系数h_a^*=1，压力角=20°。试求这对齿轮的模数m、传动比i_{12}、节圆直径及啮合角。

6. 轮系中齿轮齿数如图所示，电动机转速为1 750 r/min，计算轴1和轴2的转速。

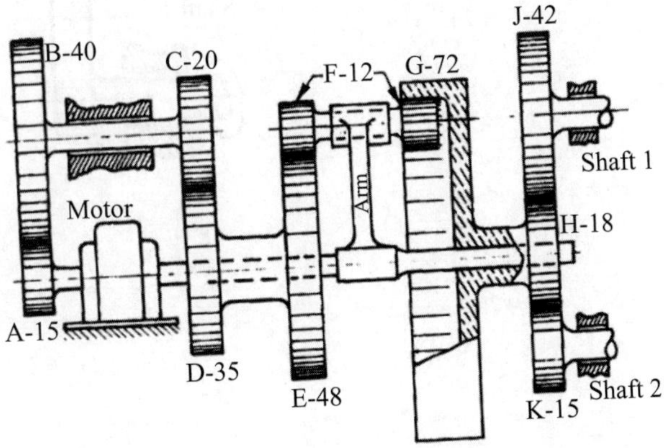

7. In the following the figure ,the teeth numbers are given: z_1=24, z_2=52, $z_{2'}$=21, z_3=78, $z_{3'}$ =18, z_4=30, z_5=78, Determine i_{1H}.

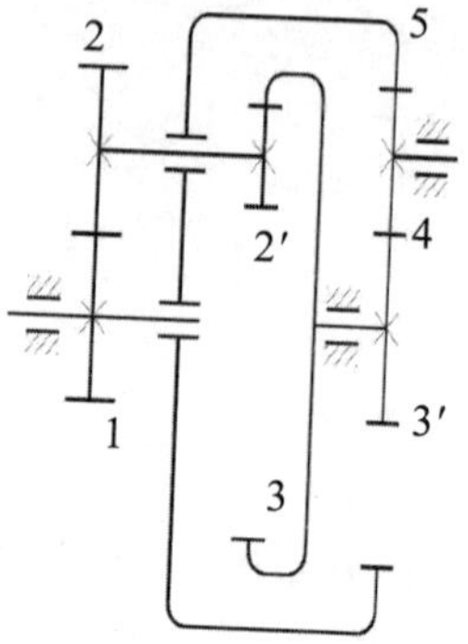

8. The teeth numbers are indicated in the figure. The arm is driven at 60 r/min. Gear A is fixed to the frame. Find the speed of gear D and direction of rotation of gear D.

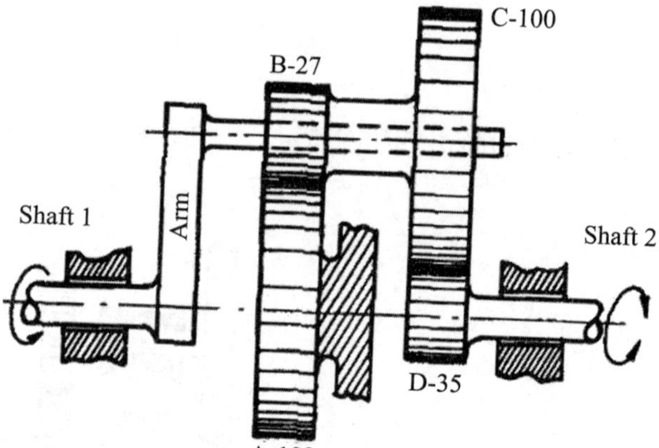

7. 在图示电动卷扬机减速器中，已知各齿轮齿数为z_1=24，z_2=52，$z_{2'}$=21，z_3=78，$z_{3'}$ =18，z_4=30，z_5=78，求i_{1H}。

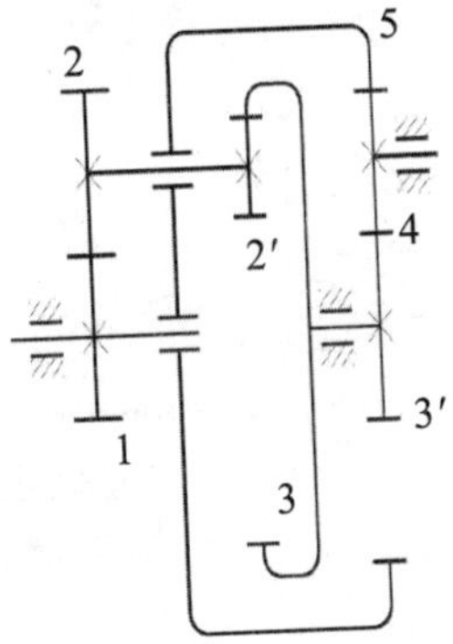

8. 轮系中齿轮齿数如图所示，系杆转速为60 r/min，齿轮A固定，求输出轴2（齿轮D）的转速和方向。

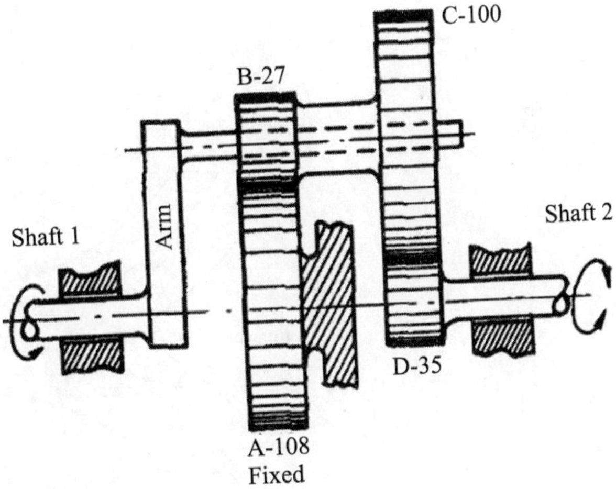

9. Shaft 1 is driven at 300 r/min, gear A is fixed to ground, the tool numbers are indicated in the figure, determine the speed and direction of shaft 2.

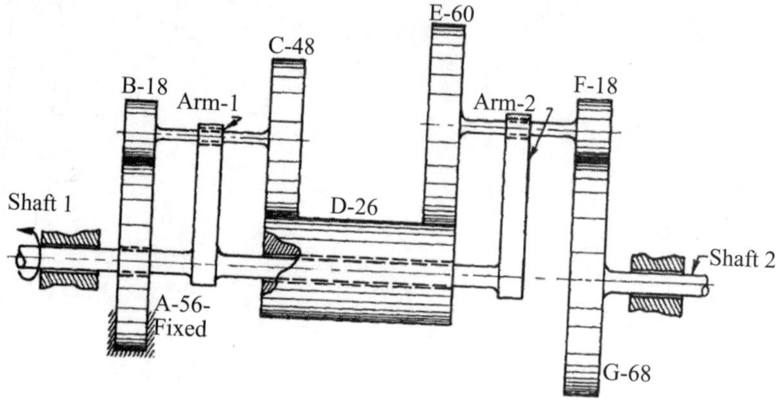

10. Explain "Interference" and it's impact.

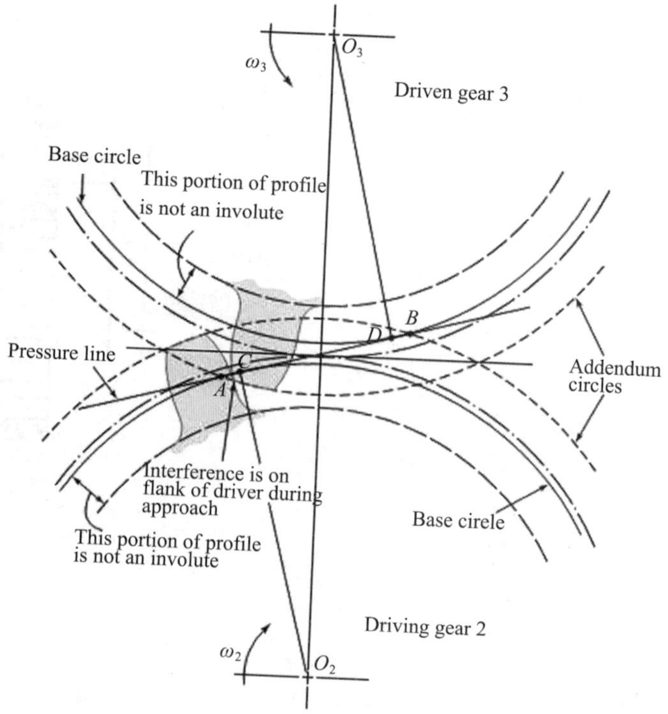

9. 轮系中齿轮齿数如图所示，轴1转动速度为300 r/min，齿轮A固定，求输出轴2的转向和速度。

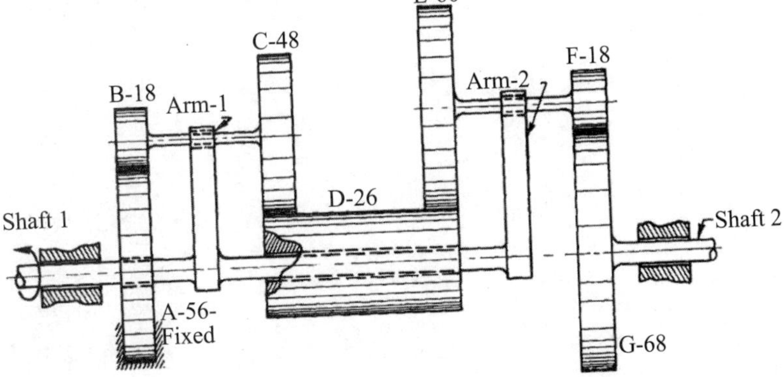

10. 解释“干涉”及其影响。

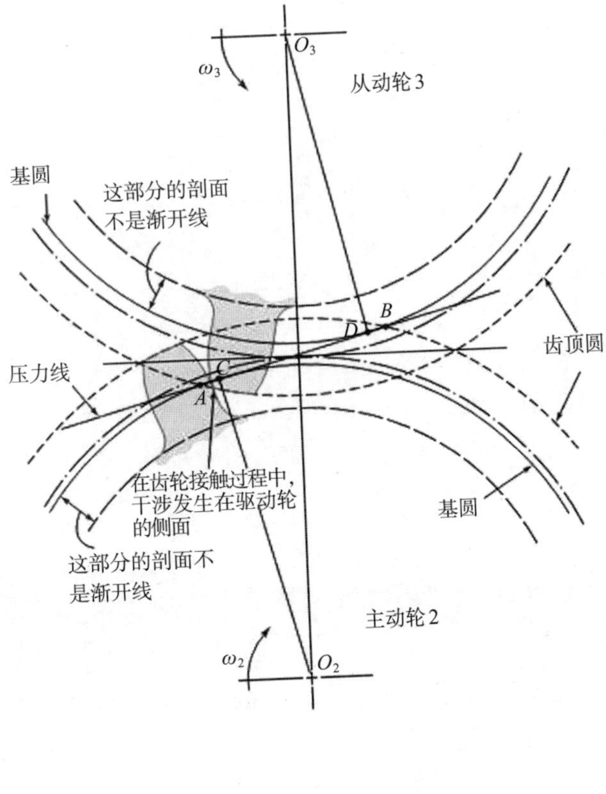

Chapter 9 Gears 2
第9章 齿 轮 2

9.1 Choice Questions

1. What is the gear failure mode targeted by the tooth surface contact fatigue strength design criterion? (   )

A. Tooth surface pitting
B. Tooth surface gluing
C. Tooth surface wear
D. Gear teeth broken

2. In gear transmission, where does the fatigue pitting of the tooth surface often occur? (   )

A. Dedendum part
B. Dedendum part close to the pitch circle
C. Addendum part
D. Addendum part close to the pitch circle

3. The main failure mode of closed hard tooth surface gear transmission is (   ).

A. tooth surface gluing
B. gear tooth fatigue fracture
C. tooth surface abrasive wear
D. gear tooth overload fracture

4. (   )of a pair of helical spur gears meshing correctly are standard values.

A. Modulus of the normal surface, the pressure angle of the normal surface on the pitch circle

B. End face modulus, end face pressure angle on the pitch circle

C. End face modulus, end face pressure angle on the pitch circle, helix angle on the pitch circle

D. Modulus of the normal surface, the pressure angle of the normal surface on the pitch circle, and the helix angle on the pitch circle

5. In a planetary gear train, any gear with a fixed geometric axis is called (   ).

A. planetary gear
B. sun gear
C. idler pulley
D. tie rod

6. A pair of standard spur gears, z_1=25, z_2=80, the bending stress is (   ).

A. $\sigma_{F1} > \sigma_{F2}$
B. $\sigma_{F1} < \sigma_{F2}$
C. $\sigma_{F1} = \sigma_{F2}$
D. $\sigma_{F1} \leqslant \sigma_{F2}$

7. When a pair of gears mesh, the contact stress values of the large and small gears is σ_{H1} (   ) σ_{H2}.

A. >
B. <
C. =

9.1 选择题

1. 齿面接触疲劳强度设计准则针对的齿轮失效形式是（　　）。

A. 齿面点蚀　　B. 齿面胶合

C. 齿面磨损　　D. 轮齿折断

2. 齿轮传动中，轮齿齿面的疲劳点蚀经常发生在（　　）。

A. 齿根部分　　B. 靠近节圆处的齿根部分

C. 齿顶部分　　D. 靠近节圆处的齿顶部分

3. 闭式硬齿面齿轮传动的主要失效形式是（　　）。

A. 齿面胶合　　B. 轮齿疲劳折断

C. 齿面磨粒磨损　　D. 轮齿过载折断

4. 一对正确啮合的斜齿圆柱齿轮传动的（　　）均为标准值。

A. 法面模数、分度圆上的法面压力角

B. 端面模数、分度圆上的端面压力角

C. 端面模数、分度圆上的端面压力角、分度圆上的螺旋角

D. 法面模数、分度圆上的法面压力角、分度圆上的螺旋角

5. 在行星轮系中，凡具有固定几何轴线的齿轮被称为（　　）。

A. 行星轮　　B. 太阳轮

C. 惰轮　　D. 系杆

6.一对标准直齿圆柱齿轮，z_1=25, z_2=80, 则这对齿轮的弯曲应力为（　　）。

A. $\sigma_{F1}>\sigma_{F2}$　　B. $\sigma_{F1}<\sigma_{F2}$

C. $\sigma_{F1}=\sigma_{F2}$　　D. $\sigma_{F1}\leqslant\sigma_{F2}$

7. 在一对齿轮啮合时，其大小齿轮接触应力值的关系是σ_{H1}（　　）σ_{H2}。

A. 大于　　B. 小于　　C. 等于

8. In cylindrical gear transmission, when the material and tooth width coefficient, tooth number ratio, load, and working conditions are constant, the contact strength of the gear mainly depends on the size of (　　).

A. modules　　　　B. number of teeth

C. pressure angle　　　　D. distance of two centers

9.2 Fill-in-the-Blanks Questions

1. When a pair of gears mesh, the relationship between the contact stress value of the large and small gears is σ_{H1}__________σ_{H2}.

2. The profile curve shape of an involute gear depends on the__________.

3. In the gear train, the gear that only changes the direction of the gear without affecting the transmission ratio of the gear train is usually called the__________.

4. Single external meshing cylindrical gear transmission, the two gears rotate in __________directions. For internal meshing cylindrical gear transmission, the two gears rotate in__________directions.

5. The following figure shows the deformation diagram of a pair of meshing gears. Please distinguish the driving wheel and the driven wheel.

Ⅰ: __________ Ⅱ:__________.

Ⅰ: concave　　　　Ⅱ: convex

8. 在圆柱齿轮传动中，材料与齿宽系数、齿数比、载荷和工作情况一定时，齿轮的接触强度主要取决于______大小。

A. 模数　　B. 齿数

C. 压力角　　D. 中心距

9.2 填空题

1. 在一对齿轮啮合时，其大小齿轮接触应力值的关系是σ_{H1}______σ_{H2}。

2. 渐开线齿轮的齿廓曲线形状取决于________。

3. 在轮系中，只改变齿轮转向而不影响轮系传动比的齿轮，通常称为______。

4. 单对外啮合圆柱齿轮传动，两齿轮的转动方向________。单对内啮合圆柱齿轮传动，两齿轮的转动方向________。

5. 如图所示为一对啮合齿轮形变示意图，请分辨其中的主动轮和从动轮。

Ⅰ：__________ Ⅱ：__________。

Ⅰ:凹面

Ⅱ:凸面

9.3 Analysis and Calculation Questions

1. For the helical cylindrical gear transmission shown in the figure, the direction of rotation of gears 3 and 4 is determined according to the principle of axial force balance of axis Ⅱ. Determine the force direction of gears 1 and 4 (three component forces marked on the diagram).

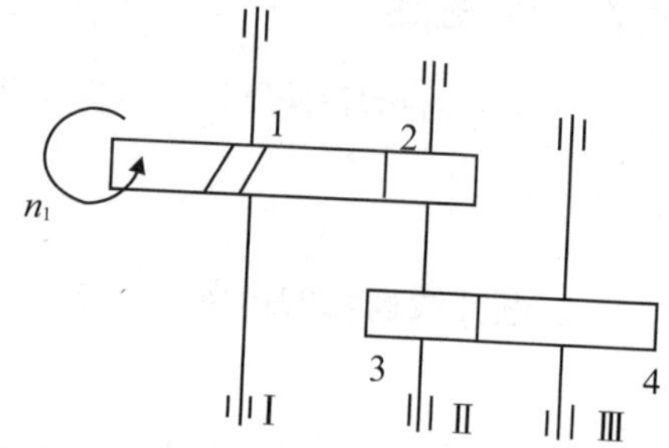

2. For the transmission shown in the figure, it is known that the number of teeth of each gear is: z_1=20, z_2=40, z_3=20, z_4=30, z_5=80, the movement are the input from the Ⅰ axis and output from the Ⅱ axis. n_1=1 000 r/min, The direction of rotation is shown in the figure. Find the rotational speed n_{II} and rotational direction of output shaft Ⅱ.

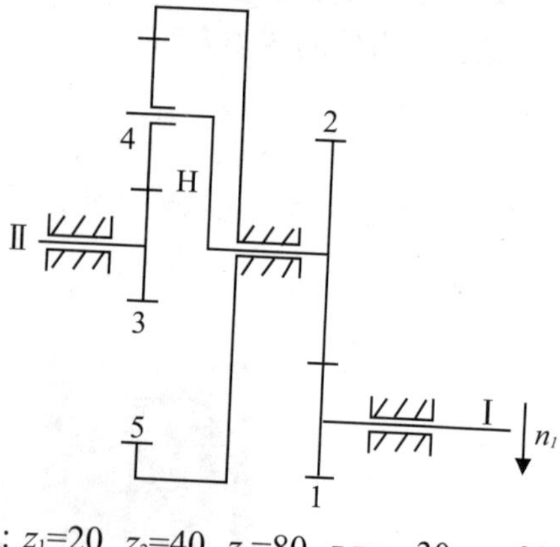

3. The gear train shown in the figure is known: z_1=20, z_2=40, z_3=80, z_4=z_5=30, z_6=90. Find i_{16}=?

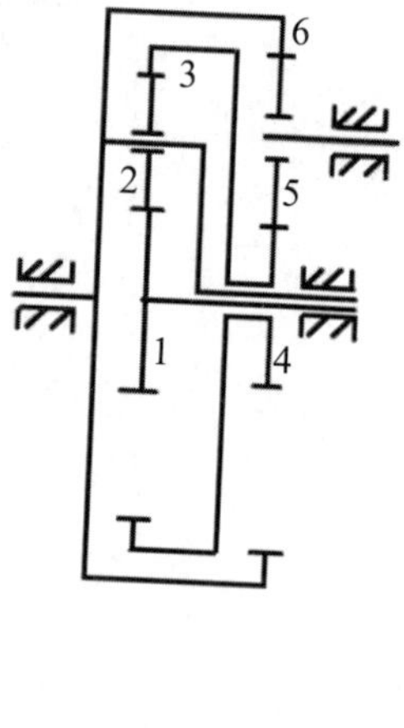

9.3 分析计算题

1. 如图所示斜齿圆柱齿轮传动，按Ⅱ轴轴向力平衡原则，确定齿轮3、4的旋向。判断齿轮1、4的受力方向（三个分力标在图上）。

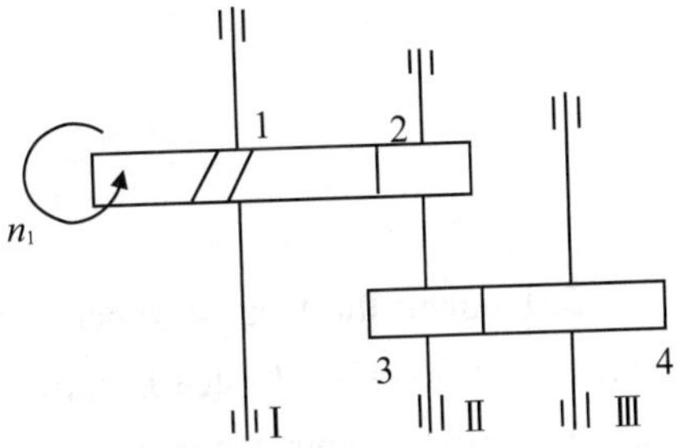

2. 在图示传动装置中，已知各轮齿数：z_1=20，z_2=40，z_3=20，z_4=30，z_5=80，运动从Ⅰ轴输入，Ⅱ轴输出，n_1=1 000 r/min，转动方向如图所示。试求输出轴Ⅱ的转速$n_{Ⅱ}$及转动方向。

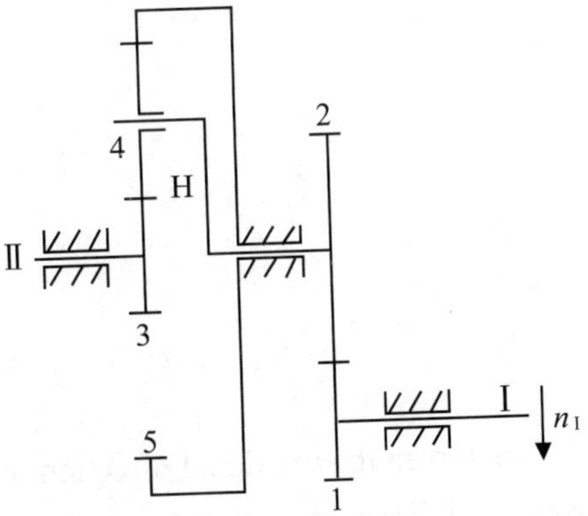

3. 如图所示轮系，已知：z_1=20，z_2=40，z_3=80，z_4=z_5=30，z_6=90。求i_{16}。

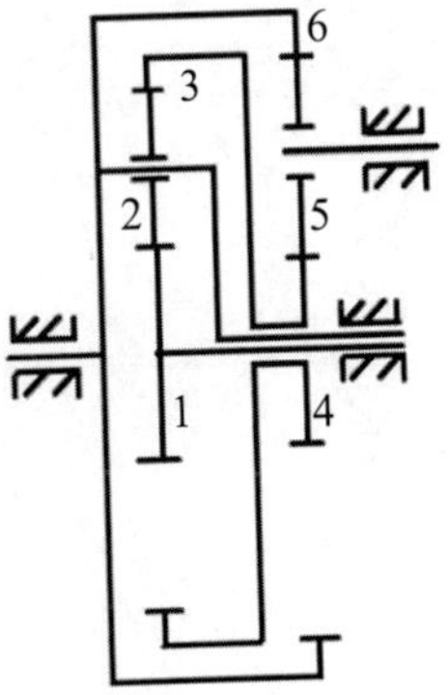

4. A standard spur gear, known pitch p=12.56 mm, number of teeth z=20, normal tooth system. Find the pitch circle diameter d, addendum circle diameter d_a, dedendum circle diameter d_f, base circle diameter d_b, tooth height h and tooth thickness s of the gear. (cos20°≈ 0.94)

5. Known: the teeth number of worm 1 z_1=1 (single-head worm), the teeth number of worm gear 2 z_2=36, the teeth number of bevel gear2′ $z_{2'}$ =24, the teeth number of bevel gear 3 z_3=30, the teeth number of gear 3′ $z_{3'}$ =20, the teeth number of internal gear 4 z_4=36 . Try to calculate the gear train transmission ratio i_{14}.

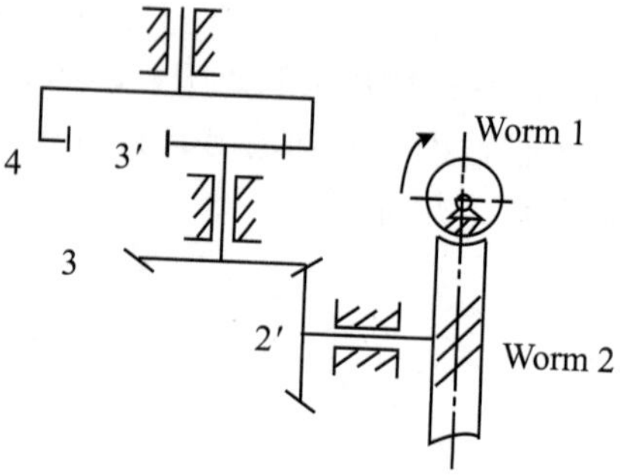

6. Given the rotational direction of head gear 1, as shown in the figure, try to judge the rotational direction of gear 4.

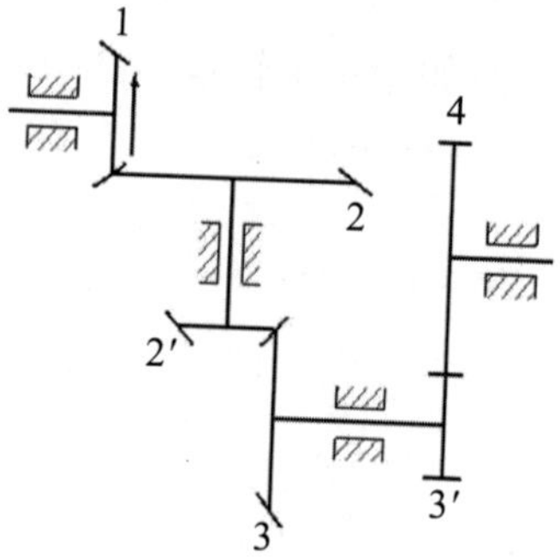

4. 某标准直齿圆柱齿轮，已知齿距p=12.56 mm，齿数z=20，正常齿制。求该齿轮的分度圆直径d、齿顶圆直径d_a、齿根圆直径d_f、基圆直径d_b、齿全高h及齿厚s。(cos20°≈0.94)

5. 已知：z_1=1（单头蜗杆），蜗轮z_2=36，锥齿轮$z_{2'}$=24，锥齿轮z_3=30，齿轮$z_{3'}$=20，内齿轮z_4=36。试计算轮系传动比i_{14}。

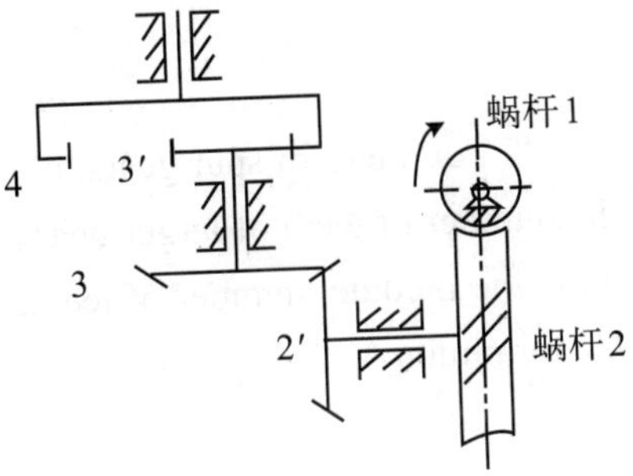

6. 已知首端齿轮1的转向，如图所示，试判断齿轮4的转向。。

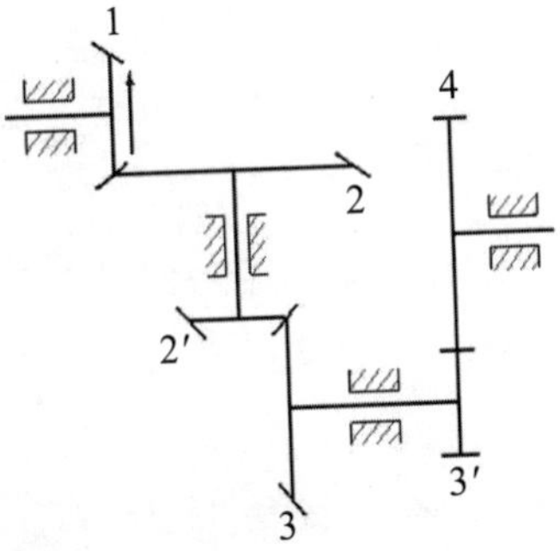

7. In the single-stage closed straight tooth standard cylindrical gear transmission, the material of the pinion is quenched and tempered with 45 steel, and the material of the big gear is ZG310-570 normalized, P = 4 kW, n_1 = 720 r/min, m = 4, z_1 = 25, z_2 = 73, b_1 = 84 mm, b_2 = 78mm. It rotates in one direction, and the load has medium impact. It is driven by motor. General reliability.

(1) Calculate the diameter of pitch circle, tooth height, diameter of addendum circle, diameter of dedendum circle, circular pitch, tooth thickness and circular pitch of base circle for small gear;

(2) Check the strength of the single-stage gear transmission.

8. For a pair of spur gears, pinion is broken. We know that center distance a=310 mm, the number of teeth of larger gear z_2=100, diameter of addendum circle d_{a2}=408 mm. Determine the module, number of teeth, diameter of pitch circle, and diameter of addendum circle for pinion.

9. The mechanism train shown consists of an assortment of gears and pulleys to drive gear 9. Pulley 2 rotates in the direction shown. Determine the direction of rotation of gear 9. Determine the hand of the helix and three forces of helical gear 6 and 7. (Attention on direction of thrust force)

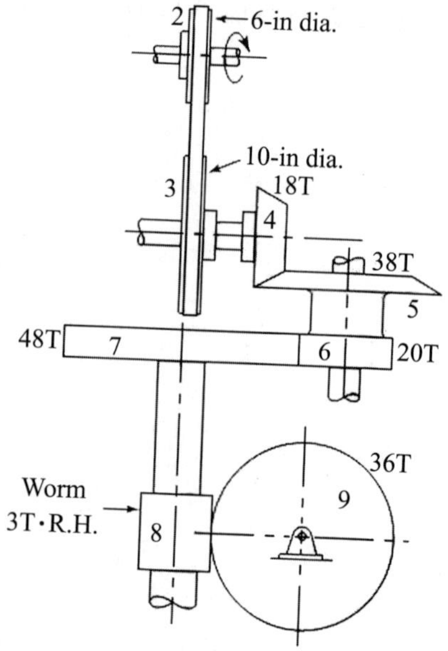

7. 在单级闭式直齿标准圆柱齿轮传动中，小齿轮的材料为45号钢调质处理，大齿轮的材料为ZG310—570正火，P=4 kW，n_1=720 r/min，m=4, z=25，z_2=73，b_1=84 mm，b_2=78 mm，单向转动，载荷有中等冲击，用电动机驱动，一般可靠性。

（1）计算小齿轮的分度圆直径、齿全高、齿顶圆直径、齿根圆直径、齿距、齿厚、基圆齿距；

（2）试验算此单级齿轮传动的强度。

8. 一对按标准中心距安装的正常齿制外啮合渐开线齿轮，中心距a=310 mm，大齿轮齿数z_2=100，齿顶圆直径d_{a2}=408 mm，试确定小齿轮的模数、齿数、分度圆直径、齿顶圆直径。

9.如图所示轮系，带轮2的旋转方向见图示。试分析齿轮9的旋转方向，斜齿轮6和斜齿轮7的旋向（注意轴向力），以及斜齿轮6和斜齿轮7的三个受力。

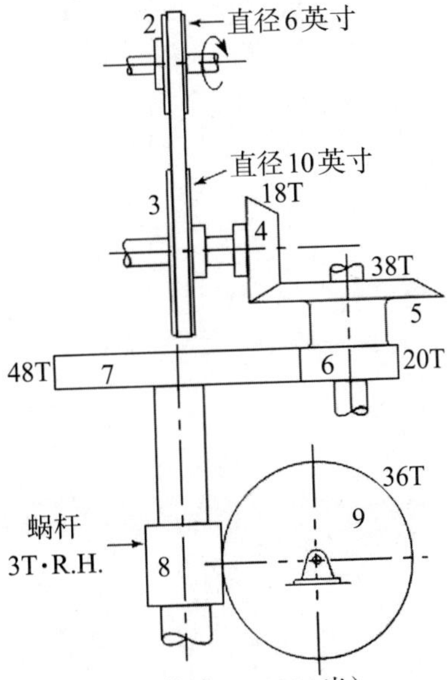

（注：1英寸≈0.0254米）

Chapter 10 Cam
第10章 凸　　轮

10.1 Choice Questions

1. When the follower of the cam mechanism adopts the law of constant velocity motion, the movement of the follower ( ).

A. will produce a rigid impact
B. will produce a flexible impact
C. no impact
D. both rigid impact and flexible impact

2. ( ) can also accurately obtain the required motion law for more complex cam profile curves.

A. The pointed follower
B. The roller follower
C. The flat-bottomed follower
D. None of the above is correct

3. ( )can make the driven follower get a larger stroke.

A. The disc cam mechanism
B. The moving cam mechanism
C. The cylindrical cam mechanism
D. None of the above is correct

4. The pressure angle of the straight-acting flat-bottom follower disc cam mechanism is ( ).

A. always equal to 0°
B. equal to a constant
C. change with the cam angle

5. For the cam mechanism with high rotational speed, in order to reduce the shock and vibration, the motion law of the follower is preferably ( ).

A. constant speed
B. constant acceleration and deceleration
C. simple harmonic motion

6. The radius of the rollers of the roller follower disc cam mechanism should be ( ) the minimum radius of curvature of the convex part of the theoretical cam profile.

A. greater than
B. smaller than
C. equal to

7. When the pressure angle of the cam increases, ( ).

A. the operation of the cam mechanism is unfavorable
B. the operation of the cam mechanism is favorable
C. the operation of the cam mechanism has no impact

10.1 选择题

1. 凸轮机构从动件选用等速运动规律时，其从动件的运动（　　）。

A. 将产生刚性冲击　　B. 将产生柔性冲击

C. 没有冲击　　D. 既有刚性冲击又有柔性冲击

2.（　　）对于较复杂的凸轮轮廓曲线，也能准确地获得所需要的运动规律。

A. 尖顶式从动杆　　B. 滚子式从动杆

C. 平底式从动杆　　D. 以上均不对

3.（　　）可使从动件得到较大的行程。

A. 盘形凸轮机构　　B. 移动凸轮机构

C. 圆柱凸轮机构　　D. 以上均不对

4. 直动平底从动件盘形凸轮机构的压力角（　　）。

A. 总是等于0°

B. 等于常数

C. 随凸轮转角而变化

5. 对于转速较高的凸轮机构，为了减小冲击和振动，从动件运动规律最好采用（　　）运动规律。

A. 等速

B. 等加速等减速

C. 正弦加速度

6. 滚子从动件盘形凸轮机构的滚子半径应（　　）凸轮理论廓线外凸部分的最小曲率半径。

A. 大于

B. 小于

C. 等于

7. 凸轮压力角增大时，（　　）。

A. 对凸轮机构的工作不利

B. 对凸轮机构的工作有利

C. 对凸轮机构的工作无影响

10.2 Fill-in-the-Blanks Questions

1. The cam mechanism is composed of three basic components: the ________, the ________, and the________ .

2. According to the shape of the cam, the cam mechanism is mainly divided into basic types such as a________ and a________ cam.

3. There are four types of contact forms between the follower and the cam profile: ________, ________, flat foot and off set.

4. The circle made with the minimum radius of the theoretical profile curve of the cam is called the________ of the cam.

5. The angle between the normal direction of a point on the cam theoretical profile curve (that is, the force direction of the driven follower) and the speed direction of the driven follower is called the________ of the cam at that point.

6. As the cam pressure angle α________, the harmful component force F_2 will increase and the follower lever will be "stuck" by self-locking. Usually, for the movable follower lever, the pressure angle α is limited during the rise.

10.3 Analysis Question

1. Design a double-dwell cam to move a follower from 0 to 6.3 cm in 60°, dwell for 120°, fall 6.3 cm in 30° and dwell for the remainder. The total cycle must take 4 s. Choose suitable programs for rise and fall. Choose radius of base circle.

2. Design a double-dwell cam to move a follower from 0 to 3.8 cm in 45°, dwell for 150°, fall 3.8 cm in 90° and dwell for the remainder. The total cycle must take 6 s. Choose suitable programs for rise and fall. Choose radius of base circle.

10.2 填空题

1. 凸轮机构是由________、________、________三个基本构件组成的。

2. 按凸轮的外形分类，凸轮机构主要分为________、________凸轮等基本类型。

3. 从动件与凸轮轮廓的接触形式有________、________、平底和球面底四种。

4. 以凸轮的理论轮廓曲线最小半径所作的圆称为凸轮的________。

5. 凸轮理论轮廓曲线上某一点的法线方向（即从动件的受力方向）与从动件速度方向之间的夹角称为凸轮在该点的________。

6. 随着凸轮压力角α________，有害分力F_2将会增大而使从动杆自锁“卡死”，通常对于移动式从动杆，推程时应限制压力角α。

10.3 分析题

1. 设计一凸轮并绘制轮廓，从动件上升位移为6.3 cm，推程运动角为60°，远休止角为120°，回程6.3 cm，回程运动角为30°，近休止角为150°。凸轮旋转一周需要4 s。自行选择从动件运动规律和基圆半径。从动件尖顶，与凸轮对心。

2. 设计一凸轮并绘制轮廓，从动件上升位移为3.8 cm，推程运动角为45°，远休止角为150°，回程3.8 cm，回程运动角为90°，近休止角为75°。凸轮旋转一周需要6 s。自行选择从动件运动规律和基圆半径。从动件滚子（半径自定），与凸轮对心。

3. A cam and roller follower in anti-clockwise turning is shown in figure, what are the three closed curves for the cam? In addition, draw up the movements of roller follower (S_1 and S_2) after cam turn in 90° and 180°.

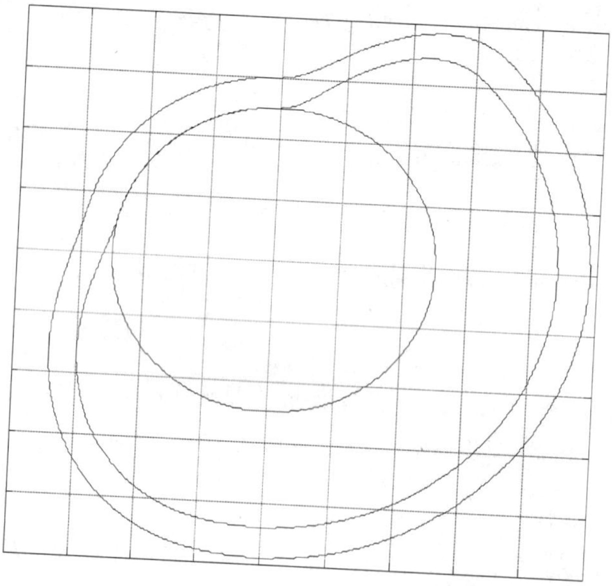

3. 如图所示从动件为滚子的凸轮机构逆时针旋转，请说明图中三条封闭的曲线分别是什么廓线。另外，在图中标出凸轮转动90°和180°时从动件走过的位移S_1（90°）和S_2（180°）,注意旋转方向。

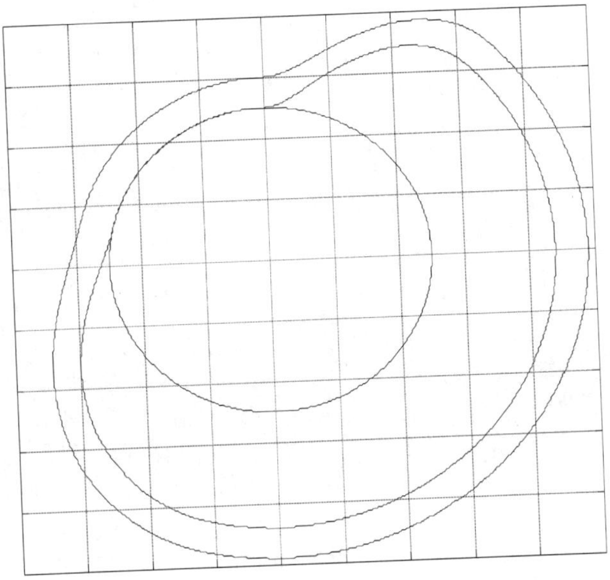

Chapter 11 Joints—Screw Thread

第11章 连接——螺纹

11.1 Choice Questions

1. From the point of view of the manufacturing process, the shaft section that needs to be threaded should have a/an ( ).

A. undercut groove
B. grinding undercut
C. chamfering
D. keyway

2. The standard profile angle for common threads is ( ).

A. 30°
B. 45°
C. 60°
D. 70°

3. When the thickness of the connected parts is not large, the threaded connections processed into through holes by the connected parts should be connected by ( ).

A. bolts
B. studs
C. screws
D. tight screws

4. Thread M24 means a common coarse thread with a ( ) of 24 mm.

A. nominal diameter
B. major diameter
C. lead
D. minor diameter

5. The connecting thread is required to be self-locking, and a ( ) thread is usually used.

A. multi-line right-handed
B. single-line right-handed
C. multi-line left-handed
D. single-line left-handed

6. The axial distance between two points corresponding to two adjacent tooth profiles of the thread is called the ( ).

A. pitch
B. lead
C. number of lines
D. nominal diameter

7. The distance the nut moves for one turn rotation is called the ( ).

A. pitch
B. lead
C. lead angle
D. nominal diameter

11.1 选择题

1. 从制造工艺出发，需切制螺纹的轴段，应留有（　　）。

A. 退刀槽　　B. 砂轮越程槽

C. 倒角　　D. 键槽

2. 普通螺纹的标准牙型角为（　　）。

A. 30°　　B. 45°

C. 60°　　D. 70°

3. 被连接件厚度不大，被连接件加工成通孔的螺纹连接应采用（　　）。

A. 螺栓连接　　B. 双头螺柱连接

C. 螺钉连接　　D. 紧钉螺钉连接

4. 螺纹M24，表示（　　）为24 mm的普通粗牙螺纹。

A. 公称直径　　B. 螺纹大径

C. 导程　　D. 螺纹小径

5. 连接螺纹要求有自锁性，通常采用的是（　　）螺纹。

A. 多线右旋　　B. 单线右旋

C. 多线左旋　　D. 单线左旋

6. 螺纹相邻两牙型对应两点的轴向距离，称为（　　）。

A. 螺距　　B. 导程

C. 线数　　D. 公称直径

7. 螺纹上任意一点沿同一条螺旋线转一周所移动的轴向距离，称为（　　）。

A. 螺距　　B. 导程

C. 升角　　D. 公称直径

11.2 Fill-in-the-Blanks Questions

1. Mechanical static connection can be divided into____________ connection and ____________connection, among which key connection, thread connection and pin connection belong to____________connection.

2. The purpose of thread connection anti-loosening is to prevent the relative movement of the thread pair. According to different working principles, there are three anti-loosening methods:____________anti-loosening, ____________anti-loosening, and other methods of anti-loosening.

3. Three-thread thread, pitch P=1.5 mm, then thread lead S=____________mm.

4. The thread angle of general pipe thread is equal to____________.

5. The section of the shaft that needs to be threaded should have an recess of thread portion, and the section of the shaft that needs to be ground should be designed with a ____________.

6. If the supporting surface of the threaded connection is uneven, it will cause additional bending stress. Therefore, ____________, countersunk seats or inclined washers should be used on the connected parts.

7. When the threaded connection is pre-tightened, the tension on the bolt is called the ____________, and the important connection needs to be strictly controlled.

11.3 Analysis and Calculation Questions

1. Why should the screw connection be pre-tightened?

11.2 填空题

1. 机械静连接又可以分为________连接和________连接，其中键连接、螺纹连接、销连接属于________连接。

2. 螺纹连接防松的目的是防止螺纹副的相对运动，按工作原理的不同有三种防松方式：________防松、________防松和其他方法防松。

3. 三线螺纹，螺距P=1.5mm，则螺纹导程S=________mm。

4. 一般管螺纹的牙型角等于________。

5、轴上需制螺纹的轴段应留有退刀槽，而需磨削的轴段应设计出________。

6. 若螺纹连接的支承面不平整，会引起附加弯曲应力，因此在被连接件上应设置________、沉头座或采用斜垫圈等。

7. 螺纹连接预紧时，螺栓所受的拉力称为________，对重要的连接需要严格控制。

11.3 分析计算题

1. 螺钉连接为什么要预紧?

2. Which design in the following is better? And give the reasons.

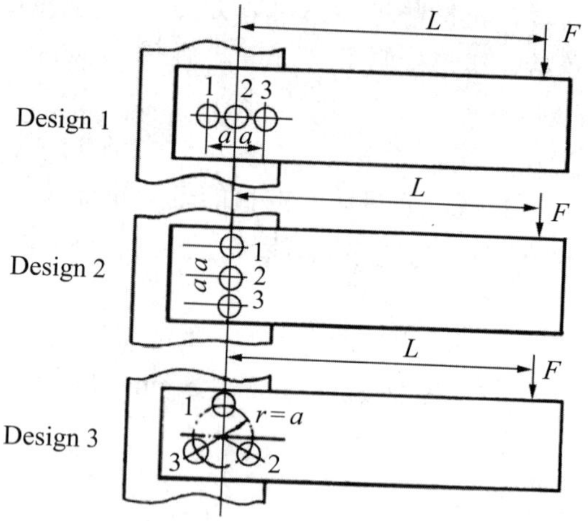

3. Bearing cover is fixed to the house by 4 bolts in the following figure. The loading working on the bearing cover $F_Q = 10.4$ kN, the material of bolt is Q235, yield strength $\sigma_s =$ 240 MPa, residual pre-tightening force $F_1 = 0.4F_Q$, pre-tightening force is not limited, safety factor S = 4, determine the minimum diameter of the bolt.

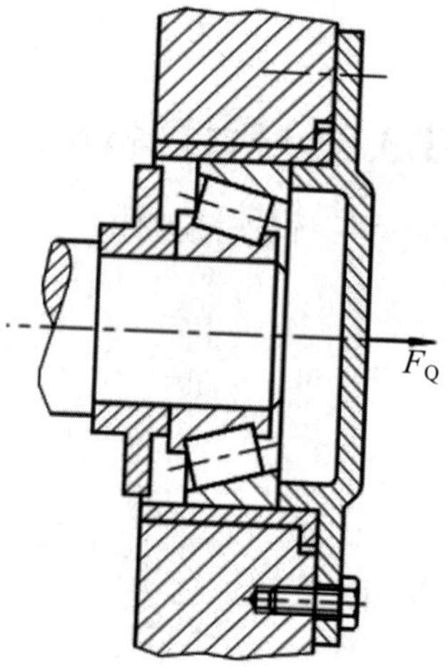

2. 试分析下图中哪个螺栓组设计方案好。

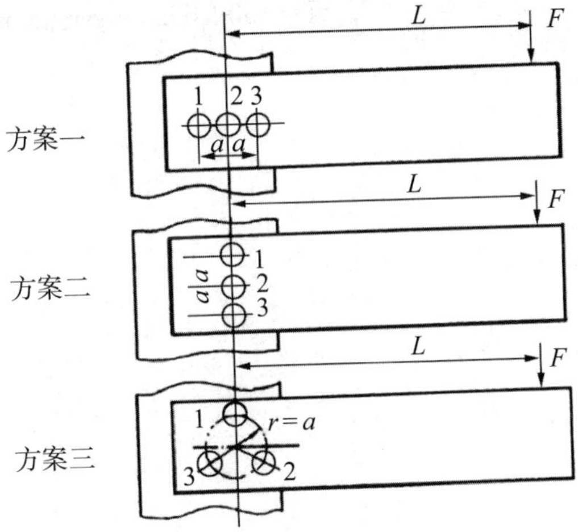

3. 图示轴承盖用4个螺钉固定于铸铁箱体上，已知作用于轴承盖上的力为F_Q = 10.4 kN，螺钉材料为Q235钢，屈服极限σ_s = 240 MPa，取残余预紧力F_1为工作拉力的0.4倍，不控制预紧力，安全系数S = 4，求螺栓所需的最小直径。

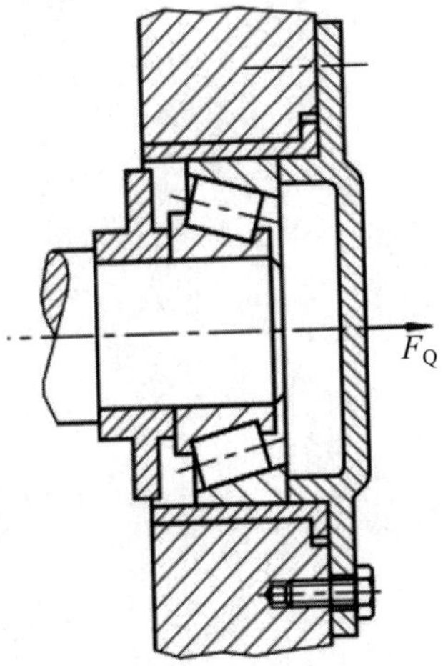

4. Two plates are jointed by two common bolts. $P = 4\ 000$ N, $L = 200$ mm, $a = 100$ mm, $f = 0.15$, $K_S = 1.2$, $[\sigma] = 300$ MPa, determine minor diameter.

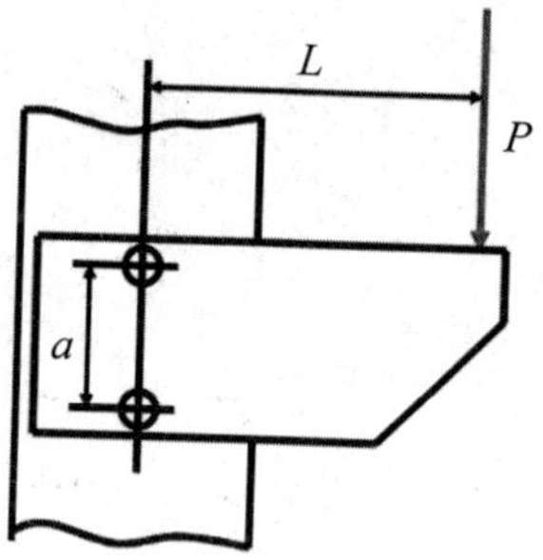

4. 竖直排列的两个普通螺栓连接两个被连接件。其中，载荷$P = 4\,000$ N，力臂$L = 200$ mm，螺栓间距$a = 100$ mm，摩擦系数$f = 0.15$，防滑系数$K_s = 1.2$，材料许用应力$[\sigma] = 300$ MPa，确定螺栓最小直径。

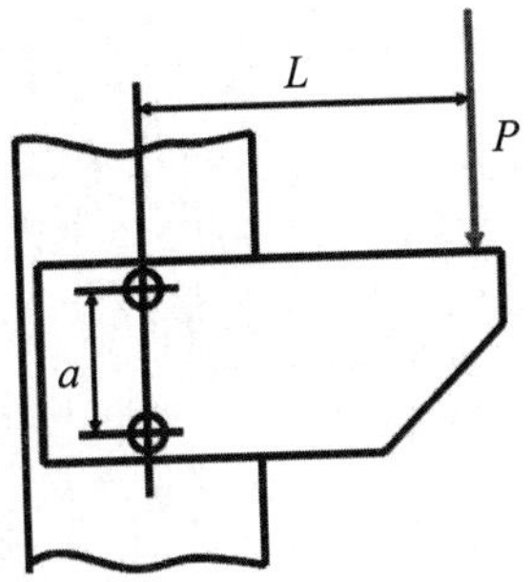

Chapter 12 Flexible Joints—Belt and Chain

第12章 柔性连接——皮带与链

12.1 Choice Questions

1. The main failure mode of the belt drive is the ( ) of the belt.

A. fatigue breaking and slipping
B. wear and gluing
C. gluing and slipping
D. wear and fatigue pitting

2. The phenomenon of creep in belt transmission occurs due to ( ).

A. the initial tension of the belt does not reach the specified value
B. the friction coefficient between the belt and the pulley is too small
C. the elastic deformation of the belt
D. the improper choice of belt type

3. ( ) is an inherent physical phenomenon in belt drives and is unavoidable.

A. Creep
B. Slipping
C. Relaxation
D. Fatigue damage

4. The more accurate transmission in belt transmission is the ( ).

A. flat belt
B. V-belt
C. round belt
D. synchronous belt

5. Flat belt transmission relies on ( ) to transmit motion.

A. the power of the driving shaft
B. the torque of the driving wheels
C. the friction between the belt and the wheels
D. None of the above

6. As shown in the figure, the correct installation position of the V-belt in the wheel groove is ( ).

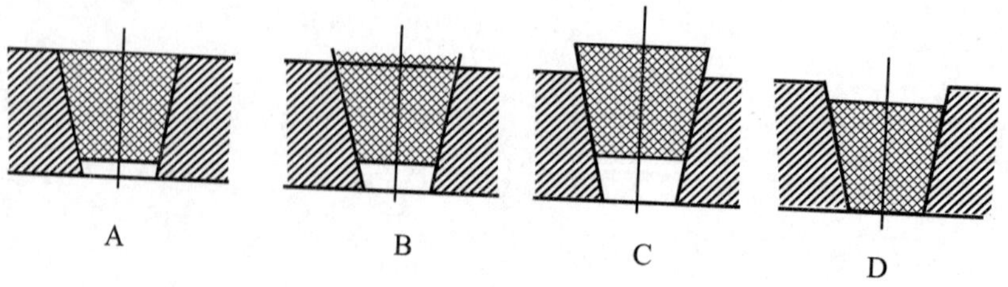

12.1 选择题

1、带传动的主要失效形式是带的（　　）。

A. 疲劳拉断和打滑　　B. 磨损和胶合

C. 胶合和打滑　　D. 磨损和疲劳点蚀

2. 带传动中出现弹性滑动现象是由于（　　）。

A. 带的初拉力达不到规定值　　B. 带与带轮的摩擦系数过小

C. 带的弹性变形　　D. 带型选择不当

3.（　　）是带传动固有的物理现象，是不可避免的。

A. 弹性滑动　　B. 打滑

C. 松驰　　D. 疲劳破坏

4. 带传动中传动比较准确的是（　　）。

A. 平带　　B. V带

C. 圆带　　D. 同步带

5. 平带传动是依靠（　　）来传递运动的。

A. 主轴的动力　　B. 主动轮的转矩

C. 带与轮之间的摩擦力　　D. 以上均不是

6. 如图所示，三角带在轮槽上正确的安装位置是（　　）。

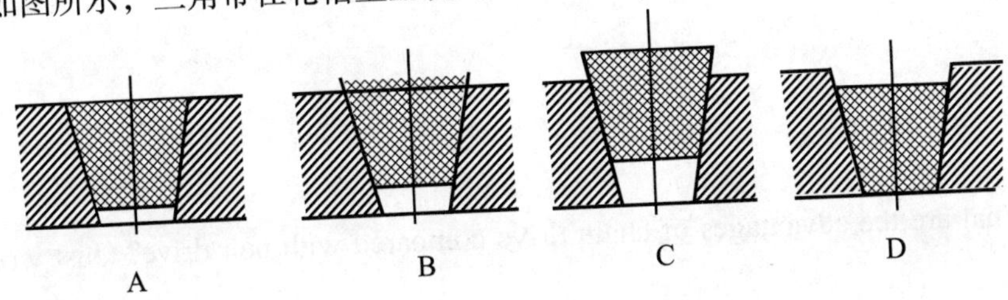

12.2 Fill-in-the-Blanks Questions

1. In order to ensure the working capacity of the belt drive, it is generally stipulated that the wrap angle of the small pulley α__________.

2. The chain drive is composed of a__________, a__________, and a__________.

3. Both chain drive and belt drive are __________parts drive.

4. The purpose of using a tensioning wheel in a belt drive is to__________.

5. Compared with gear drive and chain drive, the main advantage of belt drive is that it works__________and__________.

6. In order to make the working surfaces on both sides of the V-belt fit closely with the working surface of the wheel groove, the included angle θ of the wheel groove must be slightly__________than 40°.

12.3 Discussion Topics

1. Compared with meshing transmission such as gears, what are the advantages of belt transmission?

2. Compared with meshing transmission such as gears, what are the disadvantages of belt transmission?

3. What are the advantages of chain drive compared with belt drive? (Just write three points)

12.2 填空题

1. 为保证带传动的工作能力，一般规定小带轮的包角α________。

2. 链传动是由________、________、________所组成。

3. 链传动和带传动都属于________传动。

4. 带传动采用张紧轮的目的是________。

5. 与齿轮传动和链传动相比，带传动的主要优点是________和________。

6. 为使三角带的两侧工作面与轮槽的工作面能紧密贴合，轮槽的夹角θ______40°

12.3 讨论题

1. 与齿轮等啮合传动相比较，带传动的优点有哪些？

2. 与齿轮等啮合传动相比较，带传动的缺点有哪些？

3. 链传动与带传动相比，有哪些优点？（写三点即可）

4. Analyze stresses in the belt and determine the place of the maximum stress of belt.

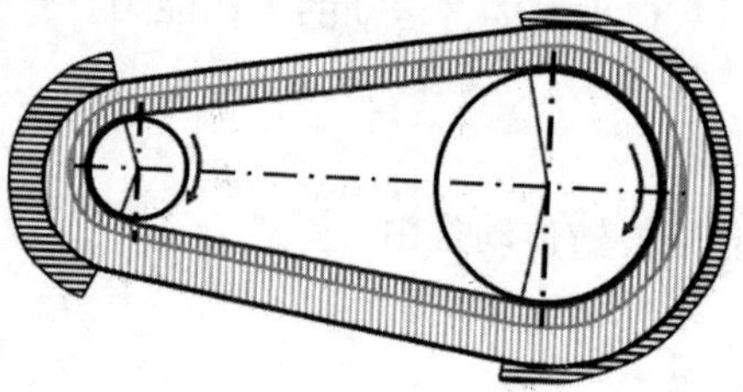

4. 分析说明如图所示的皮带传动中应力的三个组成部分（不用给出公式），并指出皮带运动过程应力最大的地方。

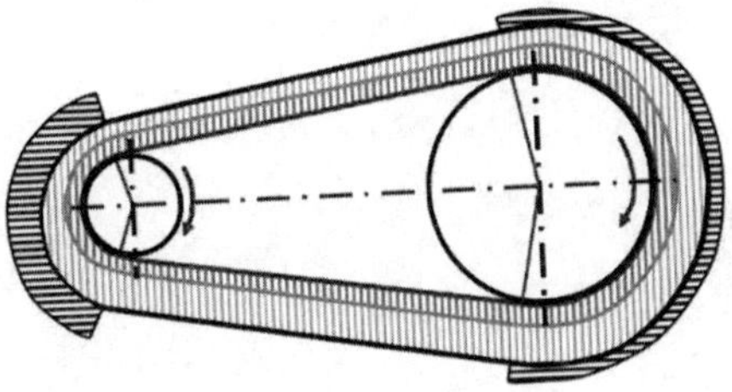

Chapter 13 Special Topics (Geometric Dimensioning and Tolerance, Tolerance and Fits and Surface Roughness)

第13章 专题（几何尺寸标注，公差与配合，表面粗糙度）

13.1 Choice Questions

1. The basis for ensuring interchangeable production is (　　).

A. generalization　　B. serialization

C. standardization　　D. optimization

2. It is the (　　) that determines the size of the dimensional tolerance zone.

A. Tolerance class　　B. Basic size

C. Actual deviation　　D. Basic deviation

3. When the measured element is a cylindrical surface, the shape of its geometric tolerance zone is the area between (　　).

A. 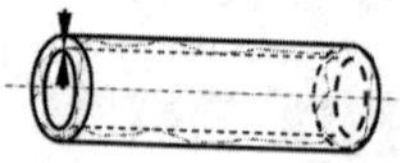

B. 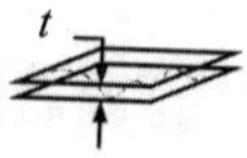

C. 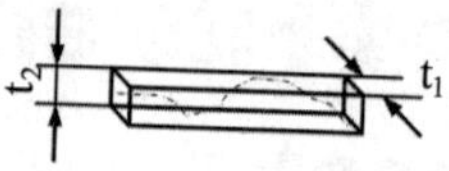

D. 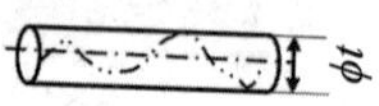

4. The surface roughness code (symbol) should be marked on (　　)on the drawing.

A. visible contour line

B. extension line

C. dotted line

D. the tip of the symbol points from the outside of the material to the marked surface

E. the tip of the symbol points from within the material to the marked surface

5. Which of the following are form tolerances? (　　)

A. Cylindricity　　B. Flatness

C. Concentricity　　D. circular runout

E. Parallelism

6. The smaller the surface roughness value, the (　　) of the parts.

A. better wear resistance

B. higher matching accuracy

C. poorer fatigue strength

D. poorer transmission sensitivity

E. easier to process

13.1　选择题

1. 保证互换性生产的基础是（　　）。

A. 通用化　　B. 系列化

C. 标准化　　D. 优化

2. 决定尺寸公差带大小的是（　　）。

A. 公差等级　　B. 基本尺寸

C. 实际偏差　　D. 基本偏差

3. 当被测要素是圆柱面时，其形位公差带的形状是（　　）之间的区域。

A. 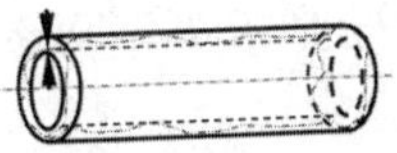

B. 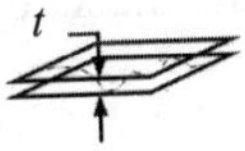

C. 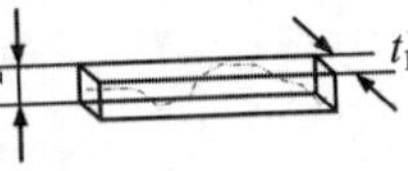

D. 

4. 表面粗糙度代（符）号在图样上应标注在（　　）。

A. 可见轮廓线上

B. 尺寸界线上

C. 虚线上

D. 符号尖端从材料外指向被标注表面

E. 符号尖端从材料内指向被标注表面

5. 下列各项中，属于形状公差的有（　　）。

A. 圆柱度

B. 平面度

C. 同轴度

D. 圆跳动

E. 平行度

6. 表面粗糙度值越小，则零件的（　　）。

A. 耐磨性越好

B. 配合精度越高

C. 抗疲劳强度越差

D. 传动灵敏性越差

E. 加工越容易

7. Which of the following statements is correct? (　　)

A. Because of mass production, parts are interchangeable, and tolerance systems are established because of interchangeable production.

B. The geometric parameters of interchangeable parts should be absolutely accurate.

C. When assembling, as long as it can be assembled without selection, it is called interchangeability.

D. A part is assembled after adjustment and passed the inspection, which is also called interchangeable production.

E. Incomplete interchange will not reduce the performance, and the economic benefits are better.

8. There are (　　) that determine the size and position of the tolerance zone.

A. standard tolerance

B. basic deviation

C. fitting tolerance

D. sum of hole and shaft tolerances

E. limit clearance or limit interference

9. The tolerance principle refers to (　　).

A. principles for determining the size of the tolerance value

B. principles for formulating tolerance and matching standards

C. principles for determining the relationship between form tolerance and position tolerance

D. principles for determining the relationship between dimensional tolerances and geometric tolerances

13.2 Fill-in-the-Blanks Questions

1. When the basic deviation code of the shaft is__________~__________, it is matched with the reference hole to form a clearance fit.

2. The surface roughness height parameters (name and code) stipulated in the national standard are__________,__________, and__________.

3. The cylindricity and radial runout tolerance zones have the same__________, but differ in__________.

7. 下列论述中，正确的有（　　）。

A. 因为有了大批量生产，所以才有零件互换性，因为有互换性生产才制定公差制。

B. 具有互换性的零件，其几何参数应是绝对准确的。

C. 在装配时，只要不需经过挑选就能装配，就称为有互换性。

D. 一个零件经过调整后再进行装配，检验合格，也称为具有互换性的生产。

E. 不完全互换不会降低使用性能，且经济效益较好。

8. 决定配合公差带大小和位置的有（　　）。

A. 标准公差

B. 基本偏差

C. 配合公差

D. 孔轴公差之和

E. 极限间隙或极限过盈

9. 公差原则是指（　　）。

A. 确定公差值大小的原则

B. 制定公差与配合标准的原则

C. 确定形状公差与位置公差的关系的原则

D. 确定尺寸公差与几何公差的关系的原则

13.2 填空题

1. 轴的基本偏差代号为________ ~ ________时，与基准孔配合形成间隙配合。

2. 国标规定的表面粗糙度高度参数（名称和代号）有________________、________________、________________。

3. 圆柱度和径向全跳动公差带的________相同，________不同。

4. Interchangeability can be divided into____________and____________.

5. The measurement process includes four elements:____________, ____________, ____________and____________.

6. The two elements of the size tolerance zone refer to the____________of the tolerance zone.

7. The smaller the value of the matching tolerance, the____________the tolerance level of the holes and shafts that cooperate with each other.

13.3 Analysis and Calculation Questions

1. Try to mark the following technical requirements on the diagram.

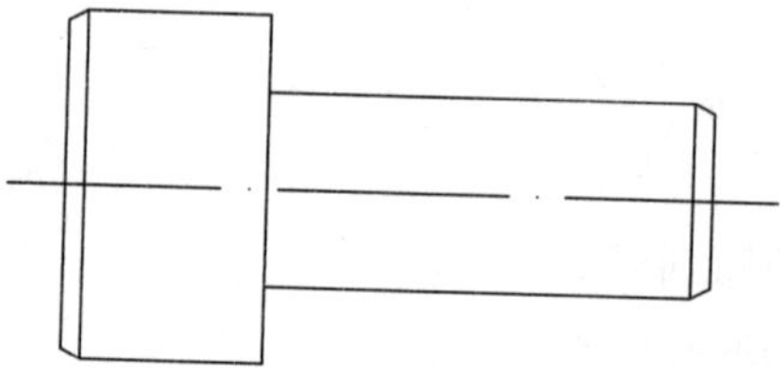

(1) The size of the cylindrical surface at the big end is required to be $\phi 45^{0}_{-0.02}$ mm , and the containment requirements shall be adopted.

(2) The concentricity tolerance of the axis of the cylindrical surface of the small end to the axis of the cylindrical surface of the large end is 0.03 mm.

(3) The size requirement of the cylindrical surface of the small end is $\phi 25^{+0.007}_{-0.007}$ mm , the straightness tolerance of the generatrix is 0.01 mm, and the tolerance requirement is adopted.

(4) The surface roughness *Ra* value of the large-end cylindrical surface is not allowed to be greater than 0.6, and the *Ra* value of the other surfaces is not allowed to be greater than 1.5.

4. 互换性可分为___________和___________。

5. 测量过程包括___________、___________、___________和___________四个要素。

6. 尺寸公差带二要素是指公差带的___________.

7. 配合公差的数值越小，则相互配合的孔、轴的公差等级愈___________。

13.3 分析计算题

1. 试将下列技术要求标注在图上。

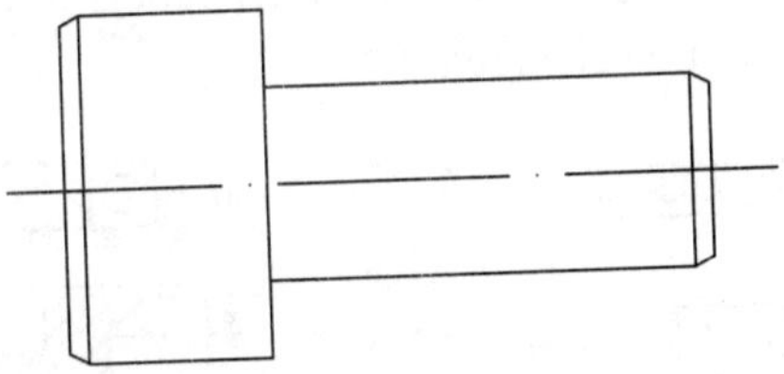

（1）大端圆柱面的尺寸要求为 $\phi 45^{0}_{-0.02}$mm，并采用包容要求。

（2）小端圆柱面轴线对大端圆柱面轴线的同轴度公差为0.03mrn。

（3）小端圆柱面的尺寸要求为 $\phi 25^{+0.007}_{-0.007}$mm，母线直线度公差为0.01mm，并采用包容要求。

（4）大端圆柱面的表面粗糙度 Ra 不允许大于0.6，其余表面粗糙度 Ra 不允许大于1.5。

2. Point out the labeling error of the geometric tolerance in the figure below and correct it. (It is not allowed to change the characteristic symbol of the geometric tolerance.)

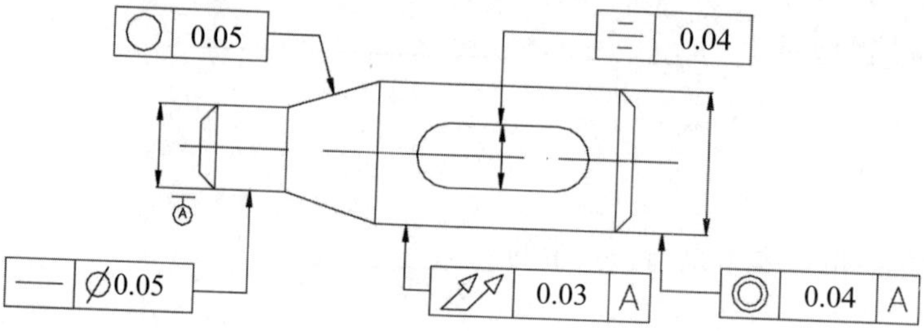

3. Correct the errors on the geometric tolerance labels in the following figures. (Geometric tolerance items must not be changed.)

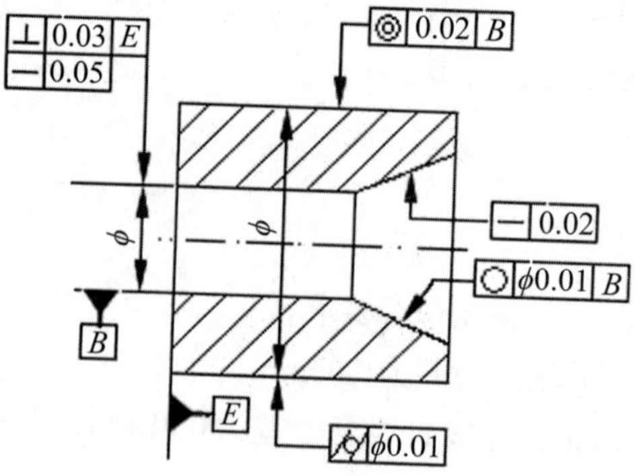

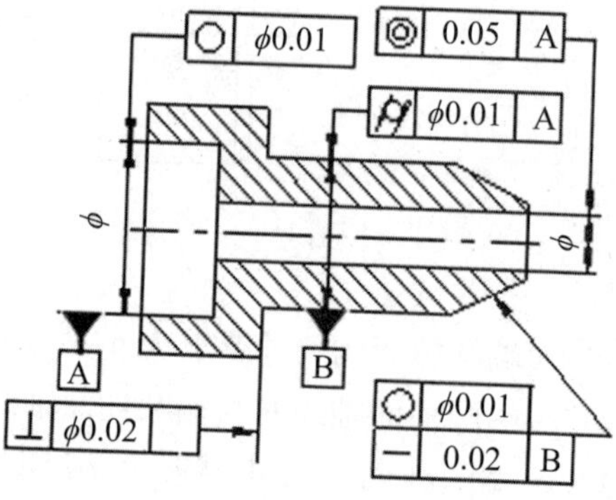

2. 指出下图中形位公差的标注错误，并加以改正。（不允许改变形位公差的特征符号）

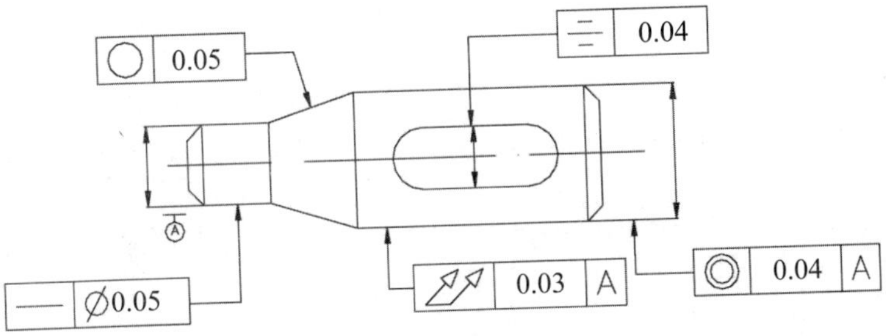

3. 改正下列图中各项几何公差标注上的错误。（不得改变几何公差项目）

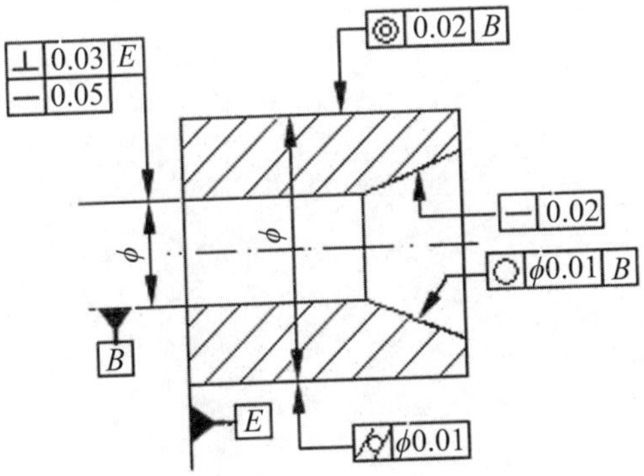

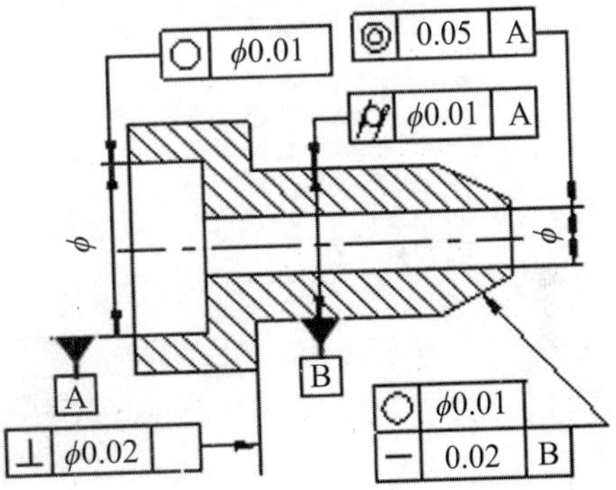

4. Mark the following geometric tolerance requirements on the drawing:

(1) The circular runout tolerance of the ϕ100h8 cylinder to the ϕ40H7 hole axis is 0.018 mm.

(2) The circular runout tolerance of the left and right bosses facing the axis of the ϕ40H7 hole is 0.012 mm.

(3) The symmetry tolerance of hub keyway to ϕ40H7 hole axis is 0.02 mm.

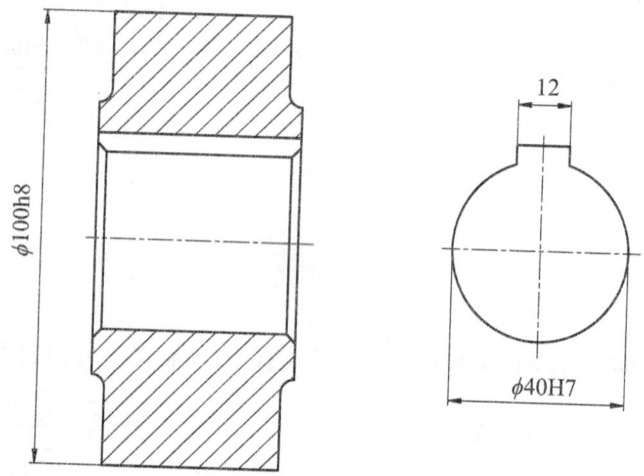

4. 将下列各项形位公差要求标注到图上：
（1）ϕ100h8 圆柱面对 ϕ40H7 孔轴线的圆跳动公差为 0.018 mm。
（2）左右两凸台端面对 ϕ40H7 孔轴线的圆跳动公差均为 0.012 mm。
（3）轮毂键槽对 ϕ40H7 孔轴线的对称度公差为 0.02 mm。

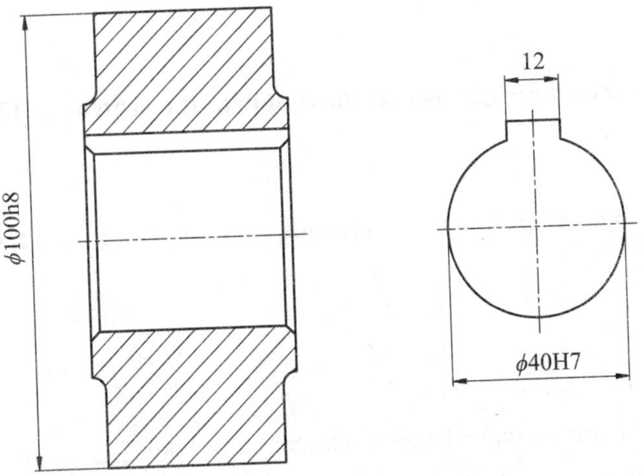

Simulated Examination Paper 1

Ⅰ. **Please select the right answer to complete the sentence.** (**12 points, 2 points per black space**)

1. Pressure angle at each point of involute curve is________; Pressure angle at the base circle is________.

A. $=$ B. $\neq$

C. $\neq 0$ D. $=0$

2.To achieve transmission between two intersecting shafts,________can be used.

A. worm drive

B. helical cylindrical gear drive

C. straight bevel gear transmission

D. straight cylindrical gear transmission

3. Compared with other mechanisms, the biggest advantage of cam mechanisms is_____.

A. to achieve various expected motion patterns

B. easy lubrication

C. easy to manufacture and achieve high precision

D. larger displacement of the follower

4. A pair of spur gear, number of teeth $z_1 < z_2$, width $b_1 > b_2$, the surface durability stress of driver gear and driven gear are s_{H1} and s_{H2}, the surface durability stress_______.

A. $\sigma_{H1} = \sigma_{H2}$

B. $\sigma_{H1} > \sigma_{H2}$

C. $\sigma_{H1} < \sigma_{H2}$

5. The life of ball rolling bearing is $L=(C/P)^{\varepsilon}$, for ball bearings, ε=_______.

A. 1 B. 3

C. 1/3 D. 10/3

Ⅱ. Please show the contact angle of bearings in the following figures. Which is the trust bearing? (8 points, 2 points per question)

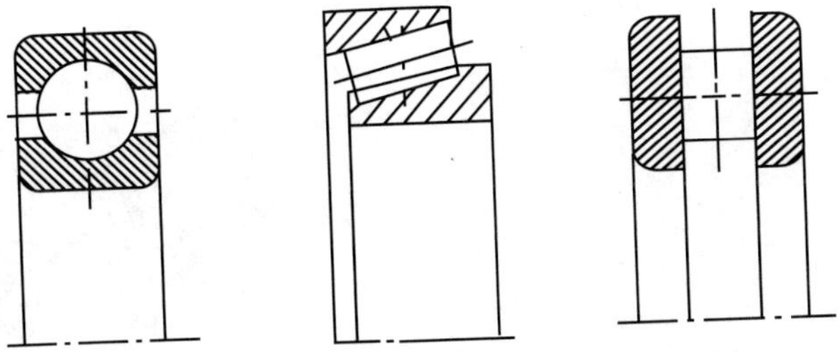

(a) (b) (c)

Ⅲ. Four-bar linkages with length of each link shown in the following figure, L_{BC} = 50 mm, L_{CD}= 35 mm, L_{AD} = 30 mm, if *AD* was ground, this mechanism is crank-rocker, link *AB* is crank, determine the maximum length of link *AB*; if this mechanism is double-crank, determine the minimum length of link *AB*. (10 points)

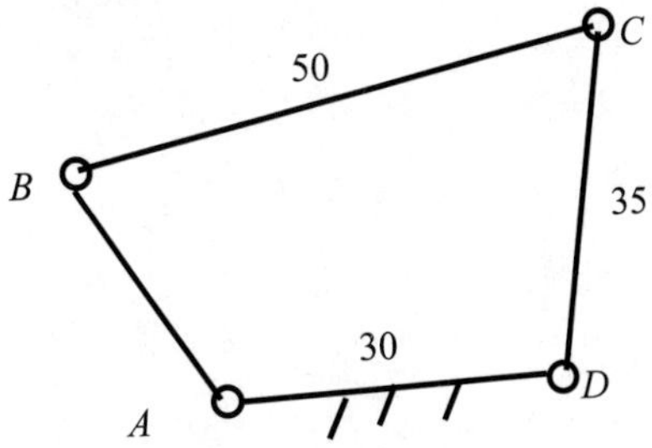

Ⅳ. Find the DOF (Degree of Freedom) of the mechanism in the following figure, give the conditions of determining motion and determine the mechanism in the following figure by conditions of determining motion. (11 points)

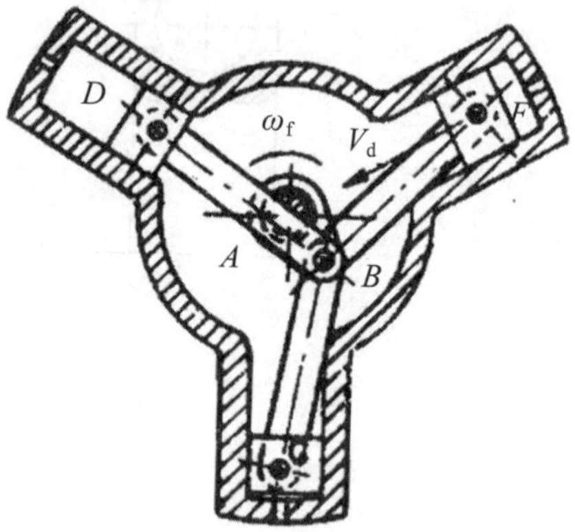

Ⅴ. For a pair of spur gears, we know that transmission ratio i=3; center distance a = 240mm and module m=5 mm. Determine diameter of addendum circle of driver gear d_{a1} and diameter of dedendum circle of driven gear d_{f2}. (10 points)

Ⅵ. Which design in the following is better? And give the reasons. (11 points)

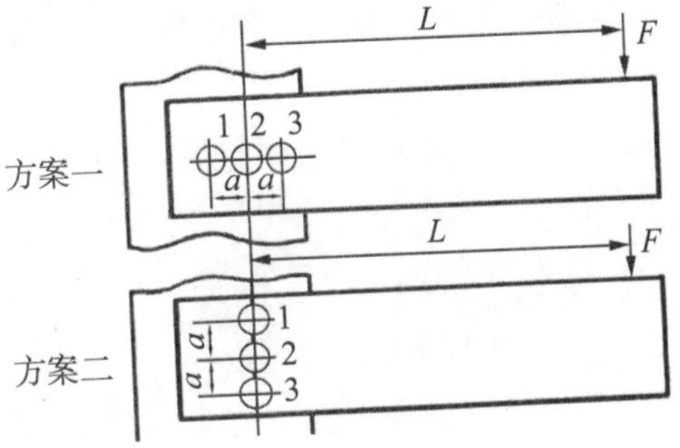

Ⅶ. Shaft 1 is driven at 300 r/min, gear A is fixed to ground, the tool numbers are indicated in the figure, determine the speed and direction of shaft 2. (11 points)

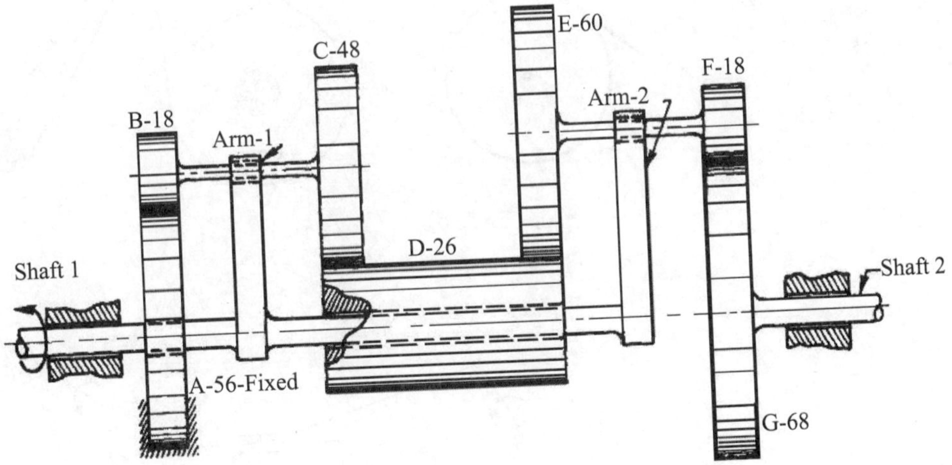

Ⅷ. Shaft a in the figure rotates in the direction shown. Find the direction of rotation of shaft d. and give three forces of helical gear 4 and 5. (8 points)

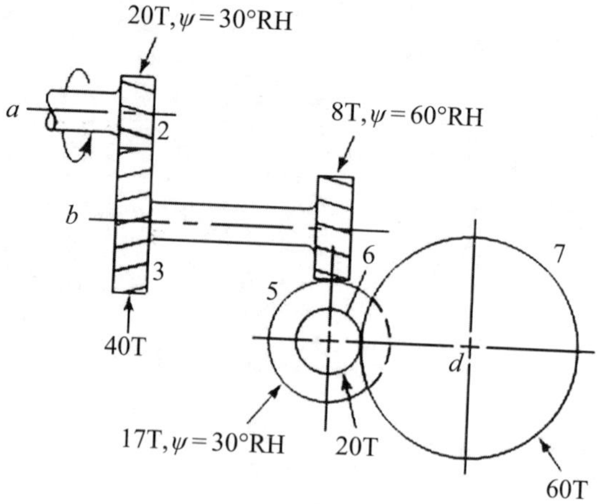

Ⅸ. Using graphical methods, find the pressure angle at the position shown. (12 points)

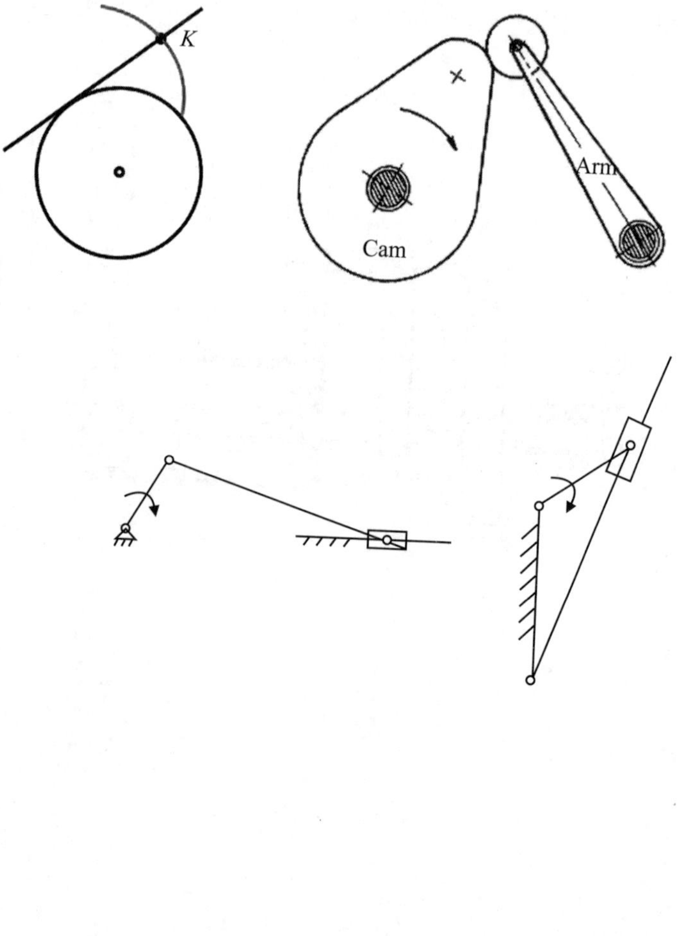

X. Please find the unreasonable designs on the shaft layout and explain the reasons. (7 points)

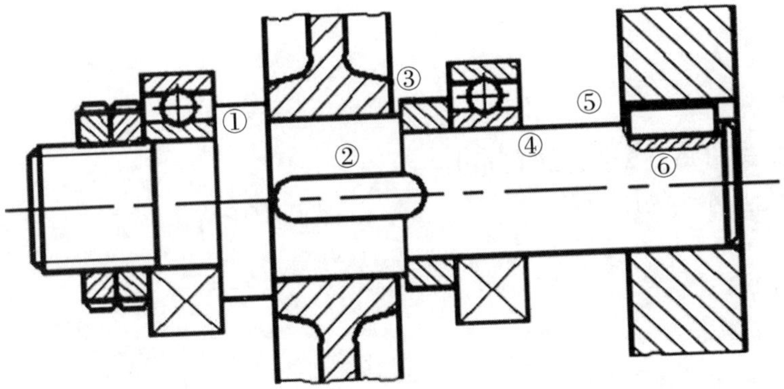

模拟试卷 1

一、选择题（每空 2 分，共 12 分）

1. 渐开线上各点的压力角__________，基圆上压力角__________。

A. 相等

B. 不相等

C. 不等于零

D. 等于零

2. 为了实现两根相交轴之间的传动，可以采用__________。

A. 蜗杆传动

B. 斜齿轮圆柱齿轮传动

C. 直齿锥齿轮传动

D. 直齿圆柱齿轮传动

3. 与其他机构相比，凸轮机构最大的优点是__________。

A. 可实现各种预期的运动规律

B. 便于润滑

C. 制造方便，易获得较高精度

D. 从动件的行程可以较大

4. 一对相啮合的直齿圆柱齿轮，齿数 $z_1 < z_2$，齿厚 $b_1 > b_2$，则这对齿轮的齿面接触应力的大小为__________。

A. $\sigma_{H1} = \sigma_{H2}$

B. $\sigma_{H1} > \sigma_{H2}$

C. $\sigma_{H1} < \sigma_{H2}$

5. 滚动轴承寿命计算公式为 $L=(C/P)^{\varepsilon}$，ε 为寿命指数，对于球轴承，ε=________。

A. 1

B. 3

C. 1/3

D. 10/3

二、请画出下图中轴承的接触角，并指出哪个为推力轴承。（每个接触角为2分，指出推力轴承2分，共8分）

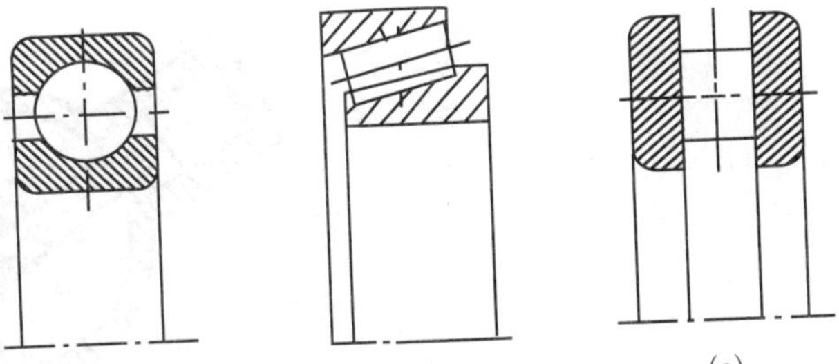

(a) (b) (c)

三、如图所示四杆结构，$L_{BC} = 50$mm，$L_{CD} = 35$mm，$L_{AD} = 30$mm，AD为机架，若此机构为曲柄摇杆机构，且AB为曲柄，求L_{AB}的最大值；若此机构为双曲柄机构，求L_{AB}的最小值。（共10分）

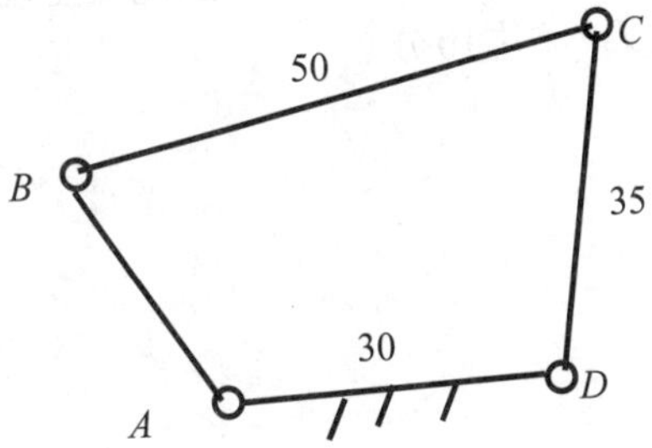

四、计算如图所示的结构自由度，驱动速度为 V_d，给出运动确定的条件，并判断运动是否确定，如有局部自由度、复合铰链和虚约束，请指出。(共11分)

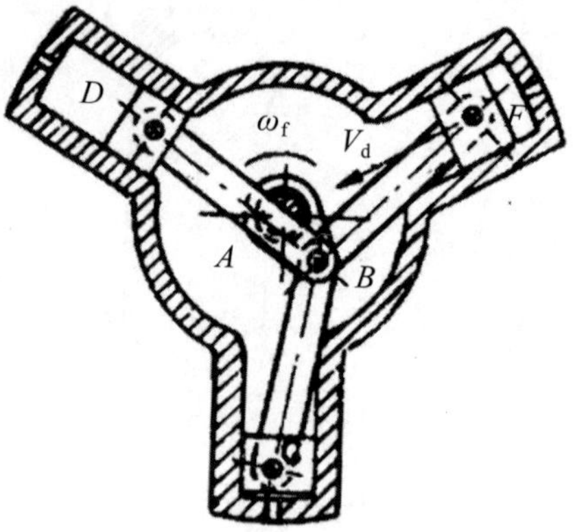

五、一对按标准中心距安装的标准直齿圆柱齿轮副，传动比为 i=3，中心距 a=240 mm，模数 m=5mm，试确定主动轮(小轮)齿顶圆直径 d_{a1}，从动轮（大轮）齿根圆直径 d_{f2}。(共10分)

六、试分析下图中哪个螺栓组设计方案好。(共11分)

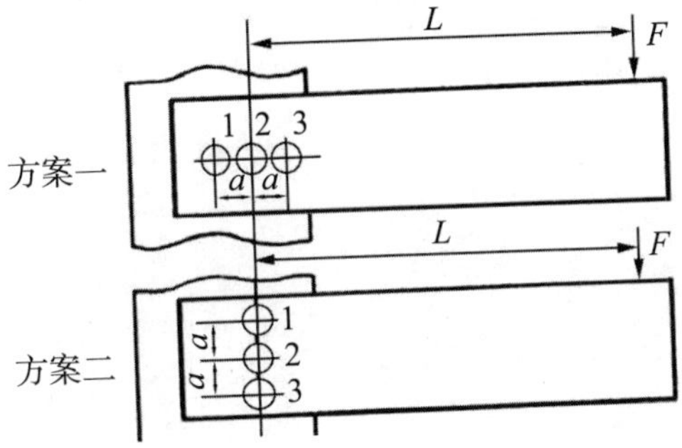

七、轮系中齿轮齿数如图所示，轴1转动速度为300 r/min，齿轮A固定，计算输出轴2的转向和速度。(共11分)

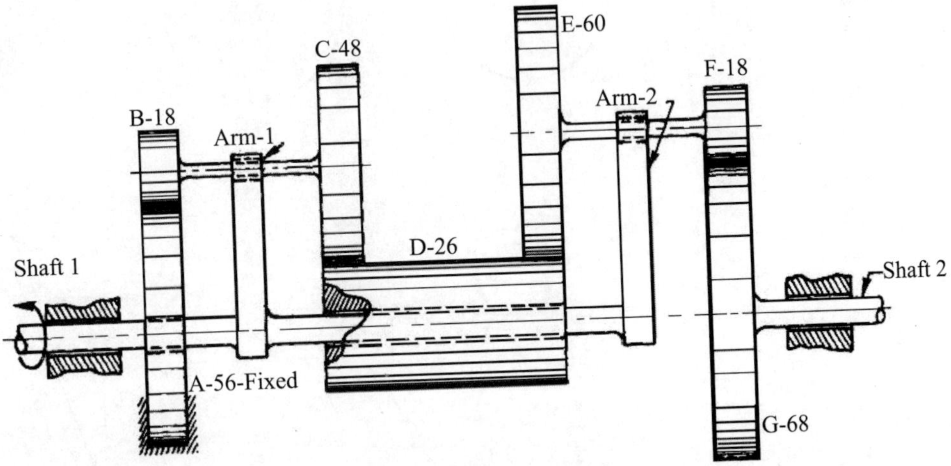

八、如图所示轮系，斜齿轮2的旋转方向见图示。试分析齿轮7的旋转方向，分析斜齿轮4和斜齿轮5的三个力。(8分)

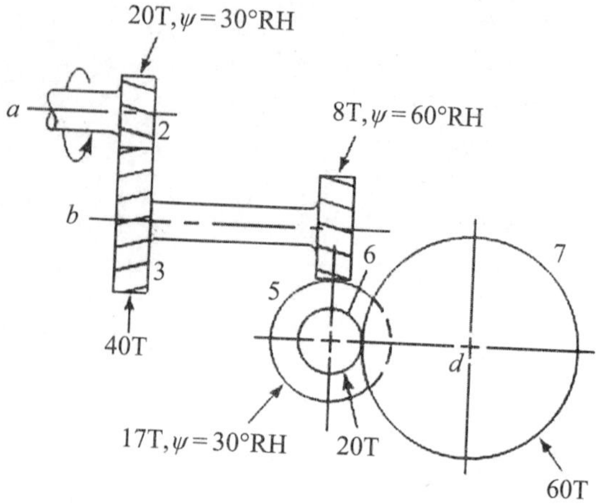

九、画出下图中的压力角。(共12分)

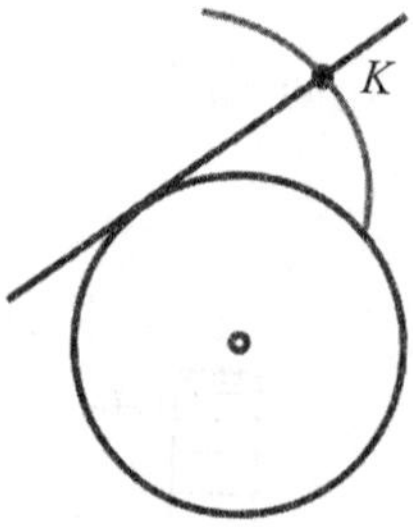

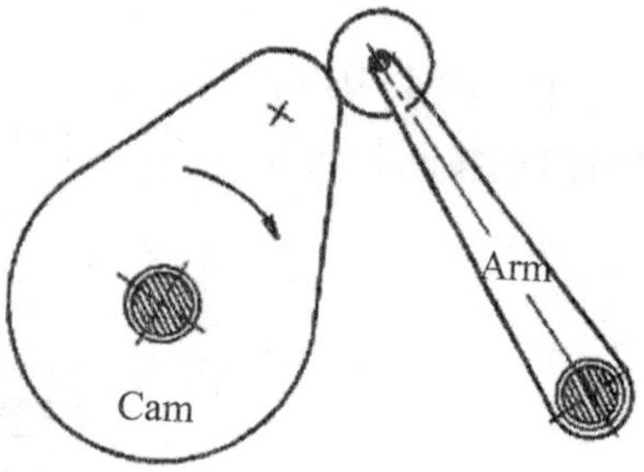

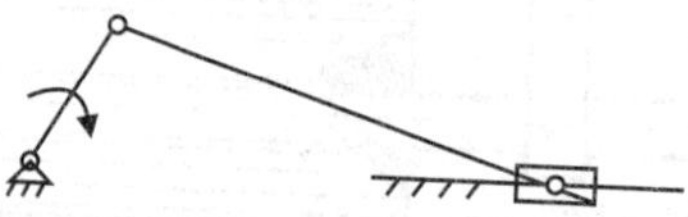

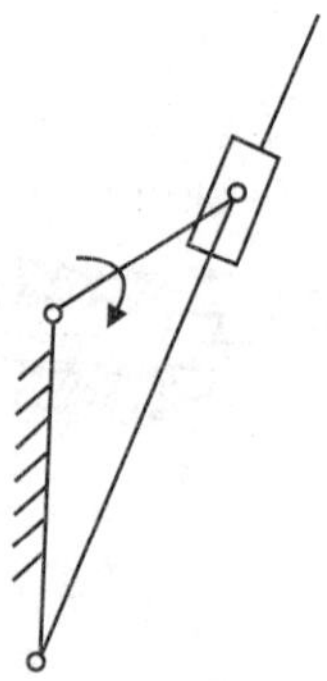

十、如下图所示，请指出并说明轴系结构中的错误。其中，轴承采用润滑油润滑，齿轮也采用润滑油润滑，直角、圆角忽略不计。不同地方的相同错误算不同错误。(共7分)

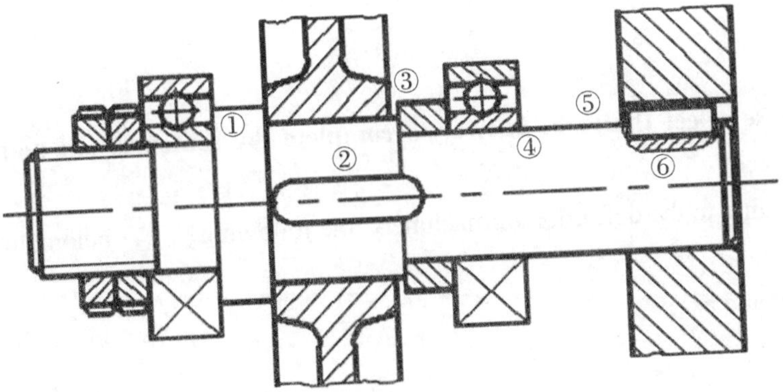

Simulated Examination Paper 2

Ⅰ. Please select the right answer to complete the sentence. (14 points, 2 points per blank space)

1. According to the definition of machinery, the following______belong to machinery.

A. lever

B. wheel and axle

C. bevel

D. gear

E. all of the above are

2. In which of the following situations, there is a locked point ? ________ .

A. Double crank mechanism

B. Centered crank slider mechanism, crank-driver

C. Crank rocker mechanism, crank-driver

D. Crank rocker mechanism, rocker-driver

3.The design criterion for tooth contact fatigue strength is aimed at the failure form of gears, which is______.

A. tooth surface pitting

B. tooth surface adhesion

C. tooth surface wear

D. tooth fracture

4. When calculating the bending strength of a gear, consider the gear as a cantilever beam and assume that all loads act______, using the root bending stress at this point as the basis for calculating the strength.

A. at the root of the tooth

B. at the pitch circle

C. at the top of the tooth

5. When one of the two connected components is too thick to be made into a through-hole and the connection does not require frequent disassembly, it is advisable to use__________ .

A. bolt connection

B. screw connection

C. stud connection

D. tight set screw connection

6. Rolling bearings 6120 has inner diameter_______ mm.

A. 20 B. 100

C. 120 D. 12

7.The rated service life of rolling bearings refers to the service life achieved when a batch of bearings of the same specification operate under specified test conditions, and the ________bearing is damaged.

A. 1% B. 5%

C. 10% D. 20%

II. Please write the right answer to complete the sentence. (17 points, 1 point per blank space)

1. When gears mesh, when the_______ of the driving gear pushes the _______ of the driven gear, a pair of teeth begin to mesh, so the starting point of meshing should be the intersection of the ______ of the driven gear and the meshing line; When the _______ of the driving gear pushes the______ of the driven gear, the two teeth are about to disengage from meshing, so the ending point of meshing is the intersection point of the______ of the driving gear and the meshing line.

2. The condition for the correct meshing of involute standard spur cylindrical gears is that the _______ and_______ are equal, respectively.

3. The commonly used machining methods for involute tooth profiles include forming cutting and______ methods.

4. The contact angle of the thrust bearing is_______ .

5. The minimum number of teeth that do not produce undercut when using rack and pinion machining for standard spur gears is _______.

6. When a pair of gears mesh, the relationship between the contact stress values of the large and small gears is σ_{H1} _______ σ_{H2}.

7. Creep is caused by the ______ during belt transmission. As long as the circumferential force is transmitted, there is a tension difference, so creep is_______. Slipping is caused by___________, so it can be___________.

8. As shown in the figure, the four link structure has L_{BC} = 50 mm, L_{CD}= 35 mm, L_{AD} = 30 mm, and AD is the frame. If this mechanism is a crank-rocker mechanism and AB is a crank, the maximum value of L_{AB} is_______________mm.

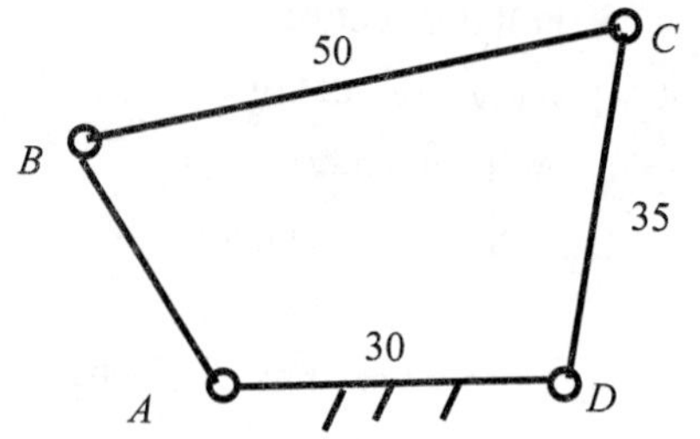

Ⅲ. Find the DOF (Degree of Freedom) of the mechanism in the following figure, give the conditions of determining motion and determine the mechanism in the following figure by conditions of determining motion. (11 points)

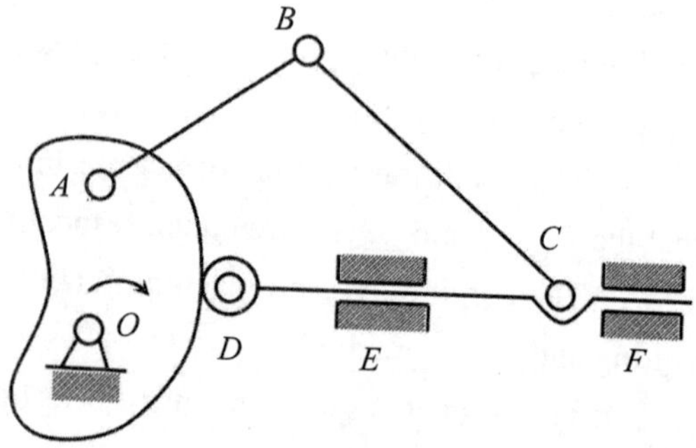

IV. For a pair of spur gears, we know that module $m = 2$ mm, numbers of teeth $z_1=25$, $z_2=75$. Determine circular pitch p, teeth height h, diameters of pitch circle d_1, d_2, diameter of addendum circle of driver gear d_{a1} and diameter of dedendum circle of driven gear d_{f2}, center distance a, transmission ratio i. (10 points)

V. Use two M10 (minor diameter d_1 = 8.376 mm, pitch diameter d_2 = 9.026 mm) screws to fix a traction hook as shown in the figure. If the screw material is Q235 steel, yield limit σ_S = 240 MPa, control the pretightening force during assembly (take the safety factor as $[S_s]$=1.6), consider the reliability coefficient of friction transmission (anti slip coefficient) K_f = 1.2, friction coefficient of joint surface μ=0.2, and calculate the allowable traction force F_R (take the calculated diameter $d_c = d_1$). (10 points)

Ⅵ. In the following the figure, the teeth numbers are given: $z_1=24$, $z_2=52$, $z_{2'}=21$, $z_3=78$, $z_{3'}=18$, $z_4=30$, $z_5=78$, determine i_{1H}. (10 points)

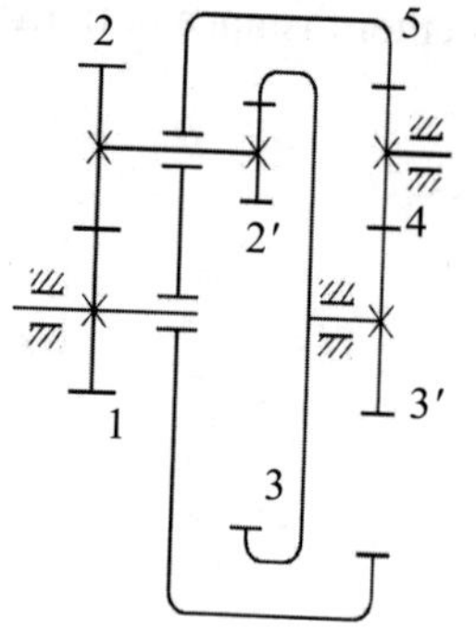

Ⅶ. Internal helical gears shown in the following figure, helical gear 1 is right hand, counterclockwise rotate.

(1) Determine the hand of the helix of helical gear 2;

(2) Give the three forces of helical gear 1 and 2 at the point P. (8 points)

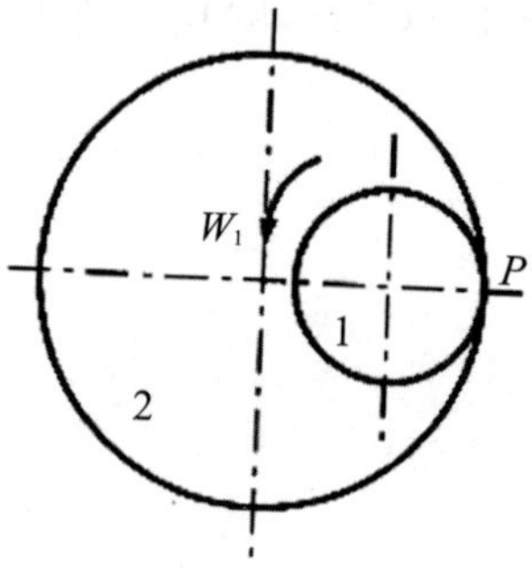

Ⅷ. A cam and roller follower in clockwise turning is shown in figure. What are the three closed curves for the cam? In addition, draw up the movements of roller follower (S_1) after cam turn in 90°. (10 points)

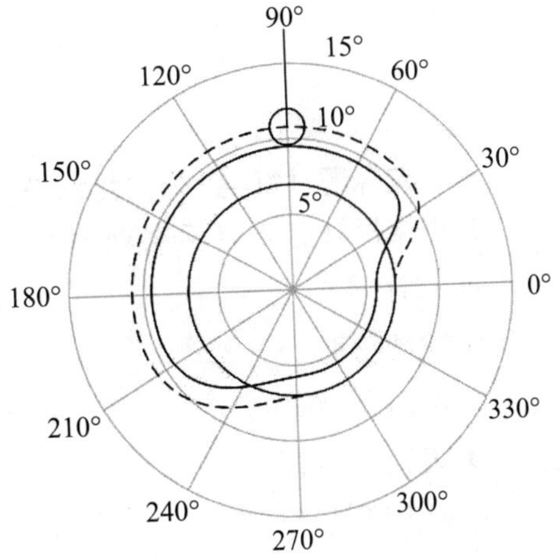

Ⅸ. Please find the unreasonable designs on the shaft layout and explain the reasons. (10 points)

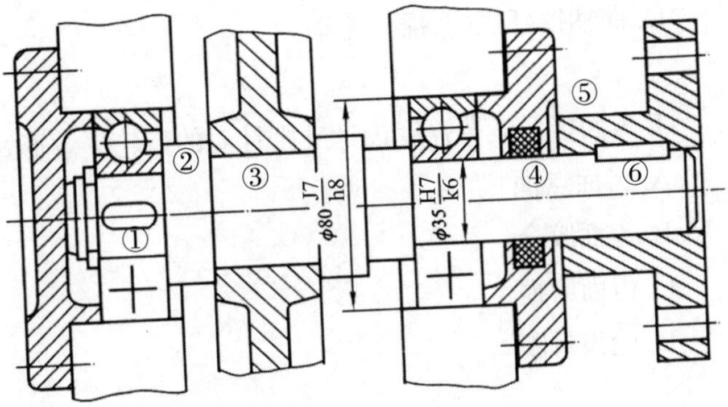

模拟试卷2

一、选择题（每空2分，共14分）

1. 根据机械定义，下列各项中属于机械的有________。

A. 杠杆

B. 轮和轮轴

C. 斜面

D. 齿轮

E. 以上都是

2. 下列各项中，存在死点的有________。

A. 双曲柄机构

B. 对心曲柄滑块机构，曲柄主动

C. 曲柄摇杆机构，曲柄主动

D. 曲柄摇杆机构，摇杆主动

3. 齿面接触疲劳强度设计准则针对的齿轮失效形式是________。

A. 齿面点蚀

B. 齿面胶合

C. 齿面磨损

D. 轮齿折断

4. 在计算齿轮的弯曲强度时，把齿轮看作一悬臂梁，并假定全部载荷作用于轮齿的________，以这时的齿根弯曲应力作为计算强度的依据。

A. 齿根处

B. 节圆处

C. 齿顶处

5.当两个被连接件之一太厚不宜制成通孔，且连接不需要经常拆装时，宜采用________。

A. 螺栓连接

B. 螺钉连接

C. 双头螺柱连接

D. 紧定螺钉连接

6. 滚动轴承6120表示内径为________mm。

A．20

B．100

C．120

D．12

7.滚动轴承的额定寿命是指一批同规格的轴承在规定的试验条件下运转，其中________的轴承发生破坏时所达到的寿命。

A．1%

B．5%

C．10%

D．20%

二、填空题（每空1分，共17分）

1. 齿轮啮合时，当主动齿轮的________推动从动齿轮的________，一对轮齿开始进入啮合，所以开始啮合点应为从动轮________与啮合线的交点；当主动齿轮的________推动从动齿轮的________，两轮齿即将脱离啮合，所以终止啮合点为主动轮________与啮合线的交点。

2. 渐开线标准直齿圆柱齿轮正确啮合的条件为________和________分别相等。

3. 渐开线齿形常用的加工方法有仿形法和________两类。

4. 推力轴承的接触角为________。

5. 标准直齿圆柱齿轮用齿条加工时不产生根切的最小齿数是________。

6. 在一对齿轮啮合时，其大小齿轮接触应力值的关系是σ_{H1} ________ σ_{H2}。

7. 弹性滑动是由于带传动时的________引起的，只要传递圆周力，就存在着拉力差，所以弹性滑动是________，而打滑是由于________引起的，所以，打滑是可以________。

8. 如图所示四杆结构，L_{BC} = 50 mm，L_{CD}= 35 mm，L_{AD} = 30 mm，*AD*为机架，若此机构为曲柄摇杆机构，且*AB*为曲柄，L_{AB}的最大值为________mm。

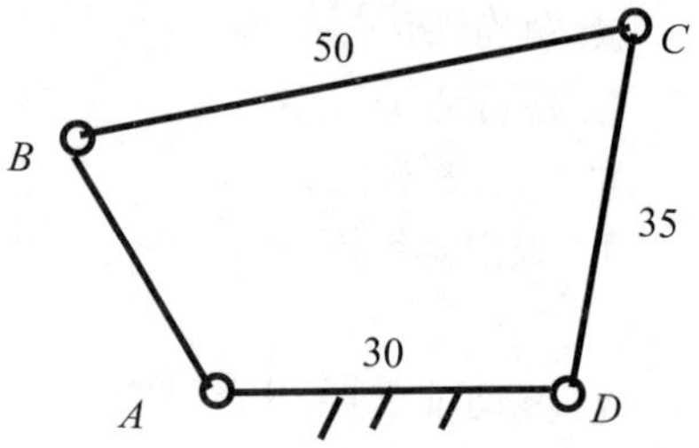

三、计算如图所示的机构自由度，给出运动确定的条件，并判断运动是否确定，如有局部自由度、复合铰链和虚约束，请指出。(共11分)

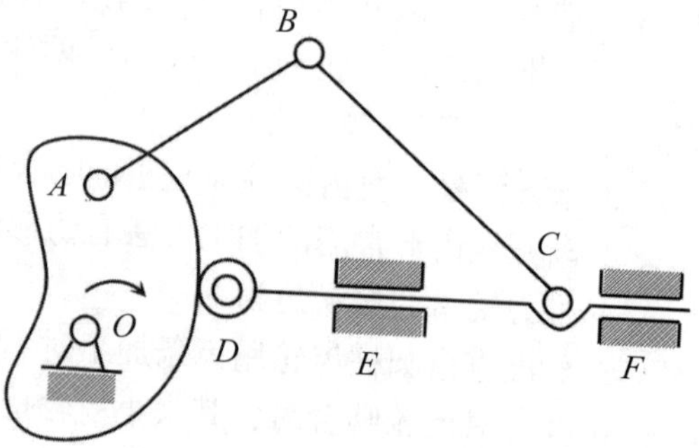

四、一对外啮合标准直齿圆柱齿轮，已知模数m=2 mm，齿数z_1=25，z_2=75。求齿距p，全齿高h，分度圆直径d_1、d_2，齿顶圆直径d_{a1}，齿根圆直径d_{f2}，标准中心距a和传动比i。（共10分）

五、如图所示，用两个M10（小径d_1 = 8.376 mm，中径d_2 = 9.026 mm，）的螺钉固定一牵引钩。若螺钉材料为Q235钢，屈服极限σ_S = 240 MPa，装配时控制预紧力（安全系数取[S_s]=1.6），考虑摩擦传力的可靠性系数（防滑系数）K_f = 1.2，接合面摩擦因数m=0.2，求其允许的牵引力F_R（取计算直径d_c=d_1，强度条件采用：1.3力/面积≤许用应力）。（10分）

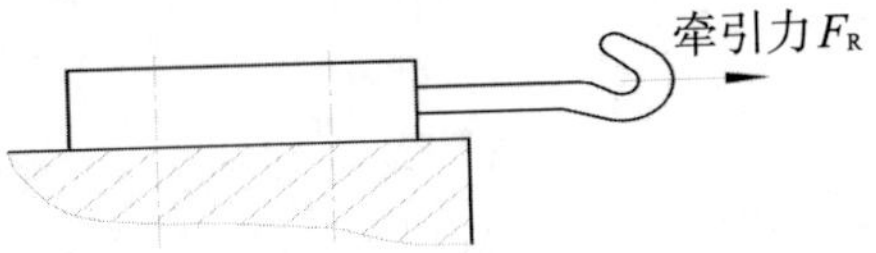

六、在图示电动卷扬机减速器中，已知各齿轮齿数为z_1=24，z_2=52，$z_{2'}$=21，z_3=78，$z_{3'}$ =18，z_4=30，z_5=78，求i_{1H}（共10分）。

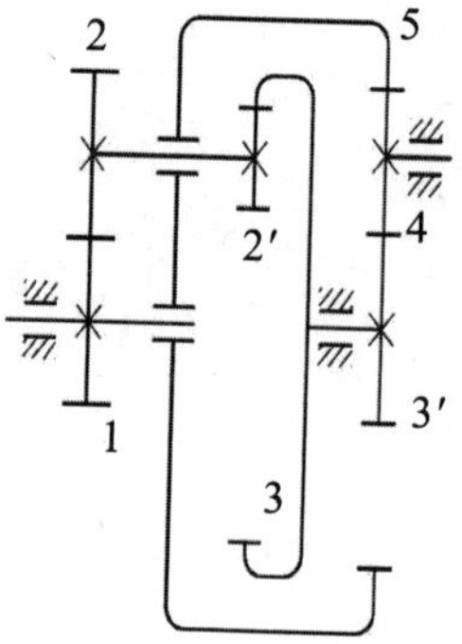

七、内斜齿轮啮合如下图所示，斜齿轮1为右旋斜齿轮，逆时针旋转。

（1）确定右旋斜齿轮2的旋向；

（2） 在点P处给出斜齿轮1和2的三个力。(共8点)

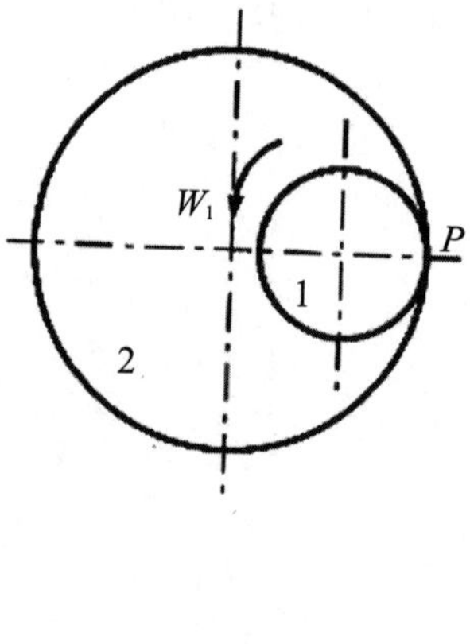

八、如图所示，从动件为滚子的凸轮机构顺时针旋转，请说明图中三条封闭的曲线分别是什么廓线。另外，在图中标出凸轮转动90°时从动件走过的位移S_1，注意旋转方向。(共10分)

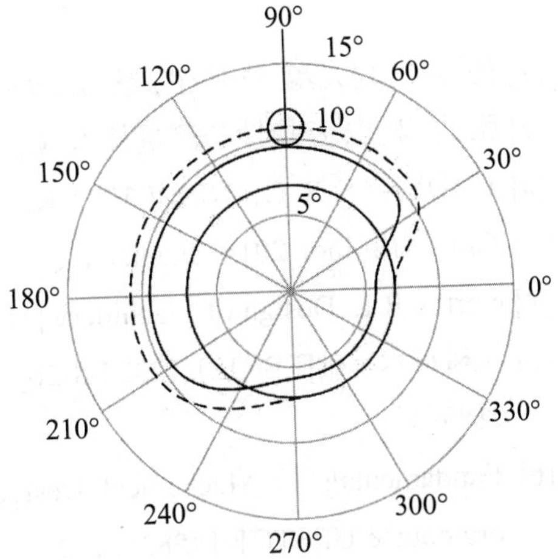

九、请指出如图所示轴系结构中的错误并给出原因，不局限于图中序号和位置，错误超过10个。(共10分)

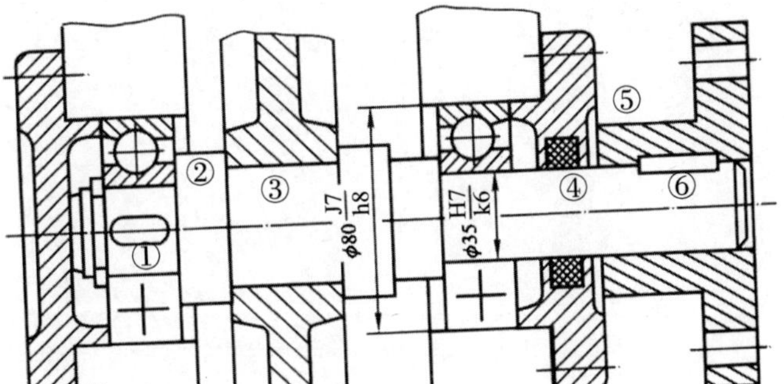

题目来源

[1] 杨可桢,程光蕴,李仲生,等. 机械设计基础[M]. 6版. 北京:高等教育出版社,2013.
[2] 张铁,李旻. 互换性与测量技术[M]. 2版. 北京:清华大学出版社,2019.
[3] BUDYNAS R G, NISBETT J K. Mechanical Engineering Design[M]. 10th ed. McGraw-Hill Inc.,2014:chapter 11.
[4] Norton R L. Design of Machinery [M]. 2nd ed.McGraw-Hill Inc.,2004:chapter 2.
[5] 机械技术基础[EB/OL]. [2024-8-20]. https://www.icourse163.org/course/UESTC-1206691838.
[6] Fundamentals of Mechanical Design. [EB/OL]. [2024-8-20].https://www.icourse163.org/course/UESTC1-1458256162.
[7] 张培培. 机械设计基础[M]. 成都:电子科技大学出版社, 2022.
[8] 百度文库. https://wenku.baidu.com.